AF581759

Symmetry-Adapted Machine Learning for Information Security

Symmetry-Adapted Machine Learning for Information Security

Editor

James (Jong Hyuk) Park

MDPI • Basel • Beijing • Wuhan • Barcelona • Belgrade • Manchester • Tokyo • Cluj • Tianjin

Editor
James (Jong Hyuk) Park
Seoul National University of Science and Technology (SeoulTech)
Korea

Editorial Office
MDPI
St. Alban-Anlage 66
4052 Basel, Switzerland

This is a reprint of articles from the Special Issue published online in the open access journal *Symmetry* (ISSN 2073-8994) (available at: https://www.mdpi.com/journal/symmetry/special_issues/Symmetry_Adapted_Machine_Learning_Information_Security).

For citation purposes, cite each article independently as indicated on the article page online and as indicated below:

LastName, A.A.; LastName, B.B.; LastName, C.C. Article Title. *Journal Name* **Year**, *Article Number*, Page Range.

ISBN 978-3-03936-642-2 (Hbk)
ISBN 978-3-03936-643-9 (PDF)

Contents

About the Editor

James (Jong Hyuk) Park received Ph.D. degrees from the Graduate School of Information Security from Korea University, Korea, and the Graduate School of Human Sciences from Waseda University, Japan. From December 2002 to July 2007, Dr. Park was a research scientist with the R&D Institute, Hanwha S&C Co., Ltd., Korea. From September 2007 to August 2009, he was a professor with the Department of Computer Science and Engineering, Kyungnam University, Korea. He is now a professor with the Department of Computer Science and Engineering and Department of Interdisciplinary Bio IT Materials, Seoul National University of Science and Technology (SeoulTech), Korea. Dr. Park has published about 300 research papers in international journals and conferences. He has served as a chair, on the program committee, or organizing committee chair for many international conferences and workshops. He is a steering chair of international conferences: MUE, FutureTech, CSA, CUTE, UCAWSN, an World IT Congress-Jeju. He is editor-in-chief of Human-centric Computing and Information Sciences (HCIS) by Springer, The Journal of Information Processing Systems (JIPS) by KIPS, and Journal of Convergence (JoC) by KIPS CSWRG. He is Associate Editor or Editor of 14 international journals including JoS, JNCA, SCN, CJ, and so on. In addition, he has been serving as a Guest Editor for international journals by some publishers: Springer, Elsevier, John Wiley, Oxford Univ. Oress, Emerald, Inderscience, and MDPI. He received best paper awards from ISA-08 and ITCS-11 conferences and outstanding leadership awards from IEEE HPCC-09, ICA3PP-10, IEE ISPA-11, PDCAT-11, and IEEE AINA-15. He received an outstanding research award from the SeoulTech in 2014. His research interests include IoT, human-centric ubiquitous computing, information security, digital forensics, vehicular cloud computing, multimedia computing, etc. He is a member of IEEE, IEEE Computer Society, and KIPS.

Editorial

Symmetry-Adapted Machine Learning for Information Security

Jong Hyuk Park

Department of Computer Science and Engineering, Seoul National University of Science and Technology (SeoulTech), 232 Gongneung-ro, Nowon-gu, Seoul 01811, Korea; jhpark1@seoultech.ac.kr

Received: 15 June 2020; Accepted: 16 June 2020; Published: 22 June 2020

Abstract: Nowadays, data security is becoming an emerging and challenging issue due to the growth in web-connected devices and significant data generation from information and communication technology (ICT) platforms. Many existing types of research from industries and academic fields have presented their methodologies for supporting defense against security threats. However, these existing approaches have failed to deal with security challenges in next-generation ICT systems due to the changing behaviors of security threats and zero-day attacks, including advanced persistent threat (APT), ransomware, and supply chain attacks. The symmetry-adapted machine-learning approach can support an effective way to deal with the dynamic nature of security attacks by the extraction and analysis of data to identify hidden patterns of data. It offers the identification of unknown and new attack patterns by extracting hidden data patterns in next-generation ICT systems. Therefore, we accepted twelve articles for this Special Issue that explore the deployment of symmetry-adapted machine learning for information security in various application areas. These areas include malware classification, intrusion detection systems, image watermarking, color image watermarking, battlefield target aggregation behavior recognition models, Internet Protocol (IP) cameras, Internet of Things (IoT) security, service function chains, indoor positioning systems, and cryptoanalysis.

Keywords: symmetry; intrusion detection system; machine learning; image watermarking; information security; IoT security; indoor positioning system

---

1. Introduction

In the current era, information, and communication technology (ICT) supports a large amount of data to provide intelligent services to next-generation industries. However, this data surge has also generated data protection challenges and created the problem of catastrophic cyberattacks. The challenges of data security are rising due to the increasing number of web-connected devices, including Internet of Things (IoT) devices and smartphones. The International Data Corporation (IDC) has forecasted that the number of connected IoT devices will reach 41.6 billion and that 79.4 zettabytes (ZB) of data will be generated by 2025. This statistic demonstrates that data growth will create significant data security problems in ICT systems. Many next-generation industries, mainly financial companies, have started investing over 10% of their total ICT budget in preventing and mitigating network and computer security threats. However, defense attempts against security threats still fail due to a lack of skilled cyber talent and the existence of low-security policies. Moreover, the changing behaviors of many security attacks, including ransomware, advanced persistent threat (APT), visualization, data compression, and supply chain attacks, result in the failure of conventional security mechanisms.

Symmetry-adapted machine-learning has shown encouraging signs of relieving security risks in ICT systems. It is a subset of artificial intelligence (AI) that relies on the principles of processing future events by learning from past events or historical data. The autonomous nature of symmetry-adapted machine-learning supports effective data processing and analysis for security detection in ICT systems without the interference of human authority. Many industries, including Amazon, Facebook, and Google, are developing machine learning-adapted solutions to support security for smart hardware, distributed computing, and the cloud. Machine learning also enables accurate product recommendations, dynamic news feeds, and smart search engines in a secure manner. Various symmetry-adapted machine-learning paradigms, including generative adversarial networks, continuous learning, one-shot learning, and deep learning have been developed to perform data processing and analysis tasks in ICT systems. These paradigms can be adapted to detect security threats, such as software exploits and unknown malware.

This Special Issue, *Symmetry-Adapted Machine Learning for Information Security*, includes the development of novel approaches with innovative architectural designs and frameworks for security attack mitigation in ICT systems by employing various machine learning paradigms. These machine learning paradigms consist of knowledge models, deep reinforcement learning, singular value decomposition, shuffled singular value decomposition, convolutional neural networks, and Q-learning. In the working period of our Special Issue, we received many submissions from many different countries, which were found to make significant contributions to the main topics of interest for the Special Issue. However, only twelve high-quality papers were accepted after three rounds of strict and rigorous review processes. Specifically, the accepted papers focus on various application areas: malware classification, intrusion detection systems, image watermarking, color image watermarking, the battlefield target aggregation behavior recognition model, IP cameras, IoT security, service function chains, indoor positioning systems, and cryptoanalysis.

2. Symmetry-Adapted Machine Learning for Information Security

Symmetry-Adapted Machine Learning for Information Security delivers successfully accepted submissions [1–12] in this Special Issue of *Symmetry*. Proposals for several innovative paradigms, novel architectural designs, and frameworks with symmetry-adapted machine learning are covered in this particular issue. All the accepted articles deliver recent developments in information security based on different dimensions: malware classification, intrusion detection systems, image watermarking, color image watermarking, the battlefield target aggregation behavior recognition model, IP cameras, IoT security, service function chains, indoor positioning systems, and cryptoanalysis.

Phuc et al. [1] propose a new attack method for the BM123-64 structure based on related-key attacks. The study addresses the security weaknesses found in ciphers based on data-dependent operations implemented in lightweight targets and rapid transformation. The study results show related-key amplified boomerang attacks on a full eight rounds of BM123-64 in distinctive designs with effective complexity results.

Kang et al. [2] present an indoor location tracking system with enhanced performance by improving unstable RSSI signals collected from BLE beacons. The error range of results obtained from the RSSI values was reduced by applying a filtering algorithm based on the average filter. The evaluation of the proposed tracking method exhibits stable performance at distances less than 7 m. However, performance degradation occurs when the distance between the beacon and the device exceeds 7 m.

The vast increase in connected devices in today's networks, and their management, information security, and complexity, are significant challenges that need to be addressed. Sun et al. [3] propose a Q-learning framework hybrid module using reinforcement learning to resolve the service chain function deployment problem in networks. Simulation-based experiment results show that the proposed algorithm is superior in performance compared to CG and Viterbi when processing service requests.

Sang et al. [4] introduce the flexible job-shop scheduling problem with parallel machines in each workstation for dynamic manufacturing systems. They propose an algorithm based on the Genetic Algorithm with two-dimensional chromosomes. Experimental analysis of the algorithm using meta-heuristic data shows an improvement of the solution by 1.34% for different dimensions of the problem, such as machine failure and bottleneck machines.

The widespread implementation of IoT devices has witnessed large-scale attacks affecting personal information leakage, Denial of Service attacks attacks, and privacy violations. Lee at al. [5] present a symmetry protocol for the efficient operation of IP cameras in the IoT environment. The authentication protocol is intended to serve in heterogeneous networks as a lightweight security solution. Performance analysis demonstrates that the protocol is lighter than existing client-side based authentication technologies, resulting in a secure and a lower energy-consuming solution.

Khan et al. [6] propose a scalable and hybrid intrusion detection system (IDS) based on a two-stage ID system using Spark machine learning and a convolutional LSTM network (Conv-LSTM). The first stage implements an anomaly detection module using Spark. The second stage performs as a misuse detection module using Conv-LSTM, addressing both global and local latent threat signatures. The IDS is evaluated using the ISCX-UNB dataset with 97.29% accuracy in identifying network misuses.

Jiang et al. [7] introduces a novel 3D-CNN model to improve the identification accuracy of battlefield target aggregation operation while maintaining the low computational cost of spatio-temporal depth neural networks. A 3D convolution two-stream model based on multi-scale feature fusion further improved the multi-fiber system reducing the computational complexity of the network. Experimental results show that the 3D-CNN model increases the efficiency of existing CNN networks for aggregate behavior recognition.

Yu et al. [8] present a robust color image watermarking algorithm for the copyright protection of color images. The algorithm is based on all phase discrete cosine biorthogonal transform (APDCBT) and shuffled singular value decomposition (SSVD). The algorithm's security and robustness are improved using SSVD and the Fibonacci transform at the watermark pre-processing stage. The experimental results show that the algorithm is resistant to attacks such as Gaussian noise, salt and pepper noise, JPEG compression, and scaling attacks.

Kim et al. [9] propose a model using multiple ε-greedy buffers in off-policy deep RL to enhance research for better generalization. Multiple random ε-greedy buffers are utilized to enhance explorations towards a near-perfect generalization. Experimental results show compatibility with discrete actions and continuous control symmetrically, resulting in improved accuracy in real-time online learning for verifying whether the network is normal or abnormal.

Khanam et al. [10] approach the challenge of proof of ownership of multimedia data exposing users to significant threats emerging from transmission channel attacks over distributed computing infrastructures. An efficient blind symmetric image watermarking method using singular value decomposition (SVD) and fast Walsh–Hadamard transform (FWHT) is discussed for ownership protection. Simulation-based results demonstrate that the proposed scheme shows high robustness against attacks with the NC being numerically one compared with existing methods, which give between 0.7991 and 0.9999.

Sarnovsky et al. [11] propose a hierarchical intrusion detection system (IDS) based on the original symmetrical combination of machine learning and a knowledge-based approach for the improved detection of new types of attacks on the network. A severity prediction model is applied, and the network operator is provided when data is low for a new attack type. The performance of the proposed knowledge-based hierarchical IDS for both precision and recall is 0.998, and 0.001 for FAR. The IDS perform better than in existing studies of network intrusion detection systems.

Kwon et al. [12] propose two malware classification methods to protect systems from attacks on their security. The symmetrical covariance matrix is used in both methods: malware classification using SimHash (MCSP), and malware classification using SimHash and linear transform (MCSLT). The performance is measured in terms of accuracy and F- score using a micro-average and a weighted macro-average. MCSP shows a maximum efficiency of 98.74% and average accuracy at 98.58% for 3-g smash encoding.

3. Conclusions

This editorial discussed information security using symmetry-adapted machine learning in ICT systems. Several mechanisms have been presented for defense against security attacks. The proposed schemes include various techniques such as malware classification, intrusion detection systems, image watermarking, and color image watermarking.

Special thanks go to the Editor-in-Chief of *Symmetry*, as well as to all the editorial teams for their invaluable support throughout the preparation and publication of this Special Issue. In addition, we thank the external reviewers for their invaluable help in reviewing the papers.

Funding: This research received no external funding.

Conflicts of Interest: The author declares no conflict of interest.

References

1. Phuc, T.; Lee, C. Cryptanalysis on SDDO-Based BM123-64 Designs Suitable for Various IoT Application Targets. *Symmetry* **2018**, *10*, 353. [CrossRef]
2. Kang, J.; Seo, J.; Won, Y. Ephemeral ID Beacon-Based Improved Indoor Positioning System. *Symmetry* **2018**, *11*, 622. [CrossRef]
3. Sun, J.; Huang, G.; Sun, G.; Yu, H.; Sangaiah, A.; Chang, V. A Q-Learning-Based Approach for Deploying Dynamic Service Function Chains. *Symmetry* **2018**, *10*, 646. [CrossRef]
4. Sangaiah, A.; Suraki, M.; Sadeghilalimi, M.; Bozorgi, S.; Hosseinabadi, A.; Wang, J. A New Meta-Heuristic Algorithm for Solving the Flexible Dynamic Job-Shop Problem with Parallel Machines. *Symmetry* **2019**, *11*, 165. [CrossRef]
5. Lee, J.; Kang, J.; Jun, M.; Han, J. Design of a Symmetry Protocol for the Efficient Operation of IP Cameras in the IoT Environment. *Symmetry* **2019**, *11*, 361. [CrossRef]
6. Khan, M.A.; Karim, M.; Kim, Y. A Scalable and Hybrid Intrusion Detection System Based on Convolutional-LSTM Network. *Symmetry* **2019**, *11*, 583. [CrossRef]
7. Jiang, H.; Pan, Y.; Zhang, J.; Yang, H. Battlefield Target Aggregation Behavior Recognition Model Based on Multi-Scale Feature Fusion. *Symmetry* **2019**, *11*, 761. [CrossRef]
8. Yu, X.; Wang, C.; Zhou, X. A Robust Color Image Watermarking Algorithm Based on APDCBT and SSVD. *Symmetry* **2019**, *11*, 1227. [CrossRef]
9. Kim, C.; Park, J. Exploration with Multiple Random ε-Buffers in Off-Policy Deep Reinforcement Learning. *Symmetry* **2019**, *11*, 1352. [CrossRef]
10. Khanam, T.; Dhar, P.K.; Kowsar, S.; Kim, J.M. SVD-Based Image Watermarking Using the Fast Walsh-Hadamard Transform, Key Mapping, and Coefficient Ordering for Ownership Protection. *Symmetry* **2020**, *12*, 52. [CrossRef]
11. Sarnovsky, M.; Paralic, J. Hierarchical intrusion detection using machine learning and knowledge model. *Symmetry* **2020**, *12*, 203. [CrossRef]
12. Kwon, Y.M.; An, J.J.; Lim, M.J.; Cho, S.; Gal, W.M. Malware Classification Using Simhash Encoding and PCA (MCSP). *Symmetry* **2020**, *12*, 830. [CrossRef]

Article

Cryptanalysis on SDDO-Based BM123-64 Designs Suitable for Various IoT Application Targets

Tran Song Dat Phuc and Changhoon Lee *

Department of Computer Science and Engineering, Seoul National University of Science and Technology, Gongneung-ro, Nowon-gu, Seoul 01811, Korea; datphuc_89@yahoo.com

* Correspondence: chlee@seoultech.ac.kr

Received: 28 June 2018; Accepted: 17 August 2018; Published: 20 August 2018

Abstract: BM123-64 block cipher, which was proposed by Minh, N.H. and Bac, D.T. in 2014, was designed for high speed communication applications factors. It was constructed in hybrid controlled substitution–permutation network (CSPN) models with two types of basic controlled elements (CE) in distinctive designs. This cipher is based on switchable data-dependent operations (SDDO) and covers dependent-operations suitable for efficient primitive approaches for cipher constructions that can generate key schedule in a simple way. The BM123-64 cipher has advantages including high applicability, flexibility, and portability with different algorithm selection for various application targets with internet of things (IoT) as well as secure protection against common types of attacks, for instance, differential attacks and linear attacks. However, in this paper, we propose methods to possibly exploit the BM123-64 structure using related-key attacks. We have constructed a high probability related-key differential characteristics (DCs) on a full eight rounds of BM123-64 cipher. The related-key amplified boomerang attack is then proposed on all three different cases of operation-specific designs with effective results in complexity of data and time consumptions. This study can be considered as the first cryptographic results on BM123-64 cipher.

Keywords: BM123-64; hybrid controlled substitution–permutation network (CSPN); switchable data-dependent operations (SDDOs); cryptanalysis; related-key amplified boomerang attack

1. Introduction

The BM123-64 [1] has a 64-bit block size covering 256-bit secret key size and a total of eight function rounds. This cipher is based on switchable data-dependent operations (SDDO) [2], which is designed to combine data-dependent operations in functions and new feature of hybrid-controlled substitution–permutation network (CSPN) models. By this way, BM123-64 is considered as a solution for a more flexible and suitable approach for appropriate application targets with each specific design. The cipher has advantages including better suitability, applicability with different algorithm designs for specific targets, and high reliability of securing against well-known attacks, for instance, linear attacks and differential attacks.

Although lots of researchers have focused on how to enhance the security of construction designs using different operations and functions, for instance, DDP (Data-Dependent Permutation) -based ciphers (such as DDP-64 [3], Cobra-family [3] and SCO (Switchable Controlled Operation) -family [2]), DDO (Data-Dependent Operation) -based ciphers (such as MD-64 [4], KT-64 [5], CTPO (Controlled Two-Place Operation) [6] and DDO-64 [7]), and SDDO-based ciphers (such as XO-64 [8] or BMD-128 [9]), their weaknesses have been recently explored with common related attacks. A simple key schedule generator for high speed transformation and lightweight targets can lead to an attack possibility for cryptanalysis using common related-key attack methods.

Related-key amplified boomerang attack [10] is an extension of the related-key boomerang attack proposed by Biham et al., 2005 [11] and Wagner, 1999 [12]. The idea of this attack is that it explores two distinctive related-key differentials to construct the related-key boomerang with high probability. Compared to other attacks, the attack was designed as an adaptive chosen plaintext attack that has become a popular and effective method to exploit many types of block ciphers. Previous studies that have applied this attack on various SDDO-based ciphers - such as COCONUT98, IDEA [11], MARS, Serpent [12], DDO-64 [13], MD-64 [14], BMD-128 [15], XO-64 [16], etc.—showed efficiency and high probabilities in cryptanalytic results.

In this paper, we propose attack methods on BM123-64 constructions with related-key approach. By constructing high probability differentials with two related-key boomerangs in distinctive designs, this attack expects to exploit a full eight rounds of BM123-64 with effective cryptanalytic results. The proposed attack requires about 2^{67} data complexity, 2^{70} memory bytes, and time complexity of 2^{67} encryptions with **Case 1** design. For **Case 2** and **Case 3** designs of BM-123-64 constructions, 2^{51} data complexity, 2^{54} memory bytes, and time complexity of 2^{65} are required. This study shows that like lots of other ciphers designed on data-dependent operations, BM123-64 still has weaknesses and is insecure against related-key cryptanalysis. The cipher construction should therefore be based on more secure primitive security approach.

The rest of this paper is organized as follows: The BM123-64 construction is briefly reviewed in Section 2. In Section 3, the proposed attacks on BM123-64 cipher is discussed, including differential characteristics (DCs), analysis methods, and security assessments. Finally, in Section 4, conclusions of the paper are presented.

2. BM123-64 Block Cipher Description

2.1. Preliminaries

This section explains notations used in this paper. With x_1 as the most significant bit and x_n as the least significant bit, a cipher X can be assigned as $X = (x_1, x_2, \ldots, x_n)$.

The DCs applied to the attack methods include descriptions of differential relation of block ciphers, such as input, output, and function round key.

- r denotes each function round of block cipher.
- $\Delta \boldsymbol{X}_{\mathbf{r}}$ denotes input difference value that occurs in each r.
- $\Delta \boldsymbol{Y}_{\mathbf{r}}$ denotes output difference value that occurs in each r.
- $\Delta \boldsymbol{U}_{\mathbf{r}}$, $\Delta \boldsymbol{Q}_{\mathbf{r}}$ denote round key difference values that occur in each r.
- e_i denotes the data bit changing within each round function, with i value considered as an active bit; at the ith position, the bit value is "1", and the remaining bits are "0" in each block data. for instance, $e_{1,3} = (1, 0, 1, 0, \ldots, 0)$).

2.2. BM123-64 Construction

BM123-64 [1] is described as a 64-bit SDDO-based block cipher with 256-bit secret key and eight function rounds in total. Each **Crypt$^{(e)}$** round function in the cipher structure does the same operation from the first round to the final round to generate output ciphertext.

The **Crypt$^{(e)}$** of BM123-64 covers three types of fixed permutation functions (I, I_1, and I_2), an extension box **E**, hybrid-controlled substitution–permutation networks **CSPNs**, and SDDO-based functions $F^{V/e}_{n/m}$ ($F_{16/64}$, $F^{-1}{}_{16/64}$, $F_{16/32}$, $F^{-1}{}_{16/32}$) based on basic controlling element $F_{2/2}$.

The process of BM123-64 encryption is described in the algorithm below.

1. 64-bit input plaintext splits into two 32-bit block A and block B.
2. From rounds r = 1 to 7, they have the same operations for each round:

   $(A, B) = \mathbf{Crypt}^{(0)}\ (A, B, U_r, Q_r)$

   $(A, B) = (B, A)$
3. In the last round, there is final transformation to output ciphertext:

$(A, B) = \mathbf{Crypt}^{(0)}\ (A, B, U_8, Q_8)$
$(A, B) = (L \oplus U_{FT}, R \oplus Q_{FT})$
$(A, B) = (A, B)$.

Figures 1 and 2 illustrate the **Crypt**$^{(0)}$ round function in detail. Reference [1] contains more description of the BM123-64 construction.

The SDDO-based functions $F_{16/64}$, $F^{-1}{}_{16/64}$ are constructed based on elementary function $F_{2/2}$, since $F_{2/2}$ is described as $((x_1, x_2), [v, z]/(y_1, y_2))$. For better performance on implementation with the target of a specific application and expansion of encryption space, the function $F^{V/e}_{n/m}$ is designed in three different operations with three different description of the basic element $F_{2/2}$.

Case 1: $y_1 = vzx_1 \oplus vz \oplus vx_1 \oplus vx_2 \oplus v \oplus z \oplus x_1 \oplus 1$
$y_2 = vzx_2 \oplus vz \oplus vx_1 \oplus vx_2 \oplus zx_1 \oplus x_2 \oplus v \oplus z \oplus 1$
$y_3 = vzx_1 \oplus vzx_2 \oplus zx_1 \oplus x_1 \oplus x_2$.

Case 2: $y_1 = vzx_1 \oplus vzx_2 \oplus vx_1 \oplus vx_2 \oplus zx_1 \oplus zx_2 \oplus z \oplus v \oplus x_2$
$y_2 = vzx_1 \oplus vzx_2 \oplus vz \oplus vx_1 \oplus vx_2 \oplus zx_1 \oplus zx_2 \oplus x_1$

$y_3 = vz \oplus v \oplus z \oplus x_1 \oplus x_2$.

Case 3: $y_1 = vx_2 \oplus x_2 \oplus x_1 \oplus v \oplus 1$
$y_2 = vx_1 \oplus x_2$.

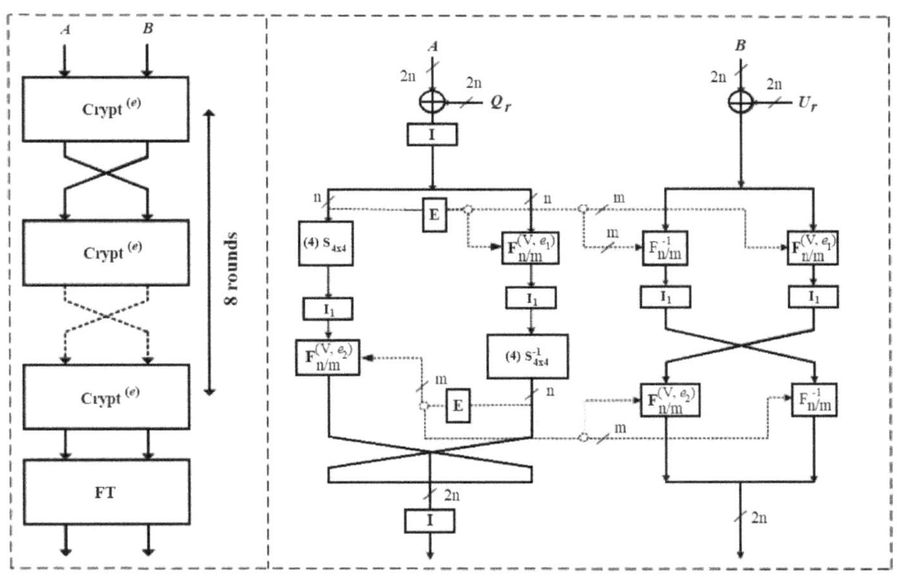

Figure 1. (**a**) The general structure (**left**); and (**b**) round function Crypt$^{(0)}$ (**right**) of BM123-64.

The three fixed permutations I, I_1, and I_2 are denoted as follows:

I_1 = (1) (2,5) (3,9) (4,13) (5,2) (6) (7,10) (8,14) (9,3) (10,7) (11) (12,15) (13,4) (14,8) (15,12) (16).
I_2 = (1) (2,3) (3,2) (4) (5) (6,7) (7,6) (8) (9) (10,11) (11,10) (12) (13) (14,15) (15,14) (16).
I = (1) (2,18) (3) (4,20) (5) (6,22) (7) (8,24) (9) (10,26) (11) (12,28) (13) (14,30) (15) (16,32) (17) (18,2) (19) (20,4) (21) (22,6) (23) (24,8) (25) (26,10) (27) (28,12) (29) (30,14) (31) (32,16).

The expansion function **E** takes a 16-bit input X, since $\mathbf{E}(X) = (X, X^{<<<4}, X^{<<<8}, X^{<<<12})$; it then outputs 64-bit controlled vector $(V, Z) = (V_1, V_2, V_3, V_4, Z_1, Z_2, Z_3, Z_4)$.

The hybrid **CSPN** construction is designed based on the description of permutation function structure covering eight 4 × 4 S-boxes (S_0, S_1, S_2, S_3 and $S^{-1}{}_0$, $S^{-1}{}_1$, $S^{-1}{}_2$, $S^{-1}{}_3$) with SDDO-based function $F_{n/m}^{V/e}$. Figure 3 presents the **CSPN** with its S-boxes in structure.

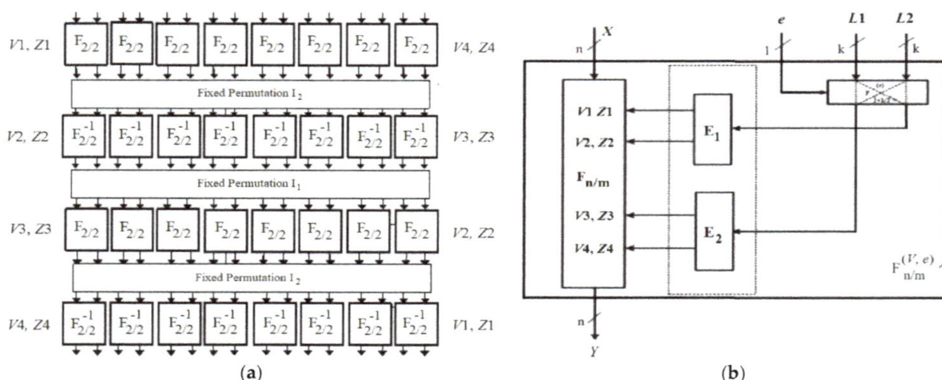

Figure 2. (**a**) Switchable data-dependent operations (SDDO)-based functions $F_{16/64}$, $F^{-1}{}_{16/64}$; and (**b**) $F_{16/32}$, $F^{-1}{}_{16/32}$.

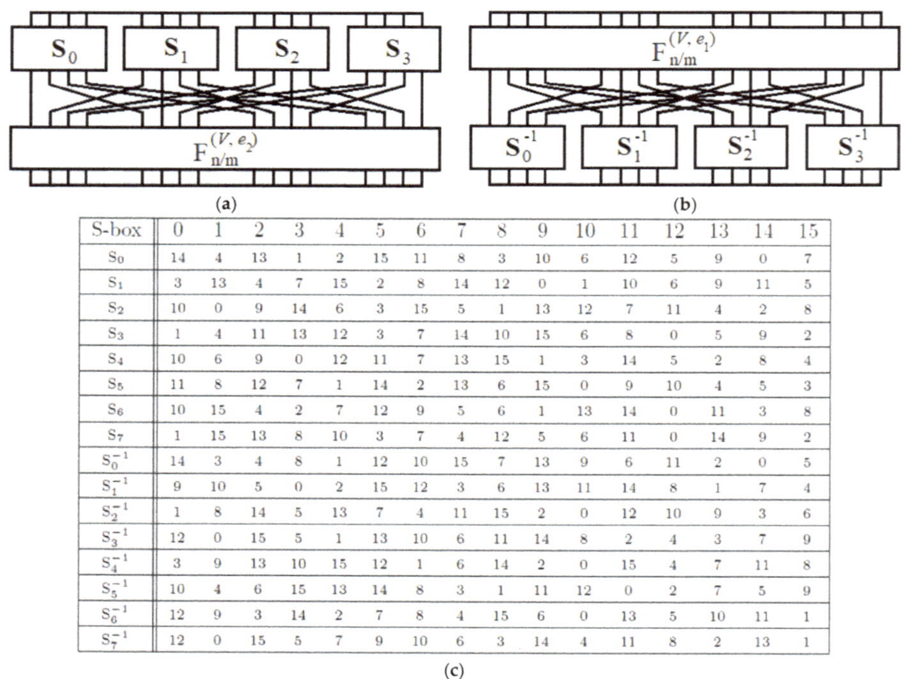

S-box	0	1	2	3	4	5	6	7	8	9	10	11	12	13	14	15
S_0	14	4	13	1	2	15	11	8	3	10	6	12	5	9	0	7
S_1	3	13	4	7	15	2	8	14	12	0	1	10	6	9	11	5
S_2	10	0	9	14	6	3	15	5	1	13	12	7	11	4	2	8
S_3	1	4	11	13	12	3	7	14	10	15	6	8	0	5	9	2
S_4	10	6	9	0	12	11	7	13	15	1	3	14	5	2	8	4
S_5	11	8	12	7	1	14	2	13	6	15	0	9	10	4	5	3
S_6	10	15	4	2	7	12	9	5	6	1	13	14	0	11	3	8
S_7	1	15	13	8	10	3	7	4	12	5	6	11	0	14	9	2
S_0^{-1}	14	3	4	8	1	12	10	15	7	13	9	6	11	2	0	5
S_1^{-1}	9	10	5	0	2	15	12	3	6	13	11	14	8	1	7	4
S_2^{-1}	1	8	14	5	13	7	4	11	15	2	0	12	10	9	3	6
S_3^{-1}	12	0	15	5	1	13	10	6	11	14	8	2	4	3	7	9
S_4^{-1}	3	9	13	10	15	12	1	6	14	2	0	15	4	7	11	8
S_5^{-1}	10	4	6	15	13	14	8	3	1	11	12	0	2	7	5	9
S_6^{-1}	12	9	3	14	2	7	8	4	15	6	0	13	5	10	11	1
S_7^{-1}	12	0	15	5	7	9	10	6	3	14	4	11	8	2	13	1

(c)

Figure 3. Controlled substitution–permutation network (CSPN) model in (**a**) left and (**b**) right of data sub-block; (**c**) different 4 × 4 S-boxes.

Like other data-dependent ciphers, BM123-64 is constructed with a very simple key schedule for high-speed transformation target. To generate function keys used in each round, 256-bit secret key K is divided into eight 32-bit subkeys $K = (K_1, K_2, \ldots, K_7, K_8)$. The key scheduling is provided with different parameters as shown in Table 1.

Table 1. Key schedule of BM123-64.

Round O_r	O_1	O_2	O_3	O_4	O_5	O_6	O_7	O_8	O_{FT}
U_r	K_3	K_4	K_8	K_6	K_2	K_7	K_5	K_2	K_3
Q_r	K_1	K_2	K_5	K_7	K_3	K_6	K_8	K_4	K_1
e'_1	1	1	1	0	0	1	1	0	-
e'_2	0	1	1	0	1	1	1	1	-
e'_3	0	0	0	1	1	0	0	1	-
e'_4	1	0	0	1	0	0	0	0	-

* O_{FT} performs the final transformation.

3. Proposed Attack Methods on BM123-64 Construction

Differential properties of operations in each round function are fundamental features to build differential characteristics and explore the related-key attack methods. Based on these properties, we can construct high probability DCs on a full eight rounds of BM123-64 with **Crypt$^{(e)}$** function. Furthermore, the related-key amplified boomerang attacks will be addressed with effective complexity results.

3.1. BM123-64 *Crypt*$^{(e)}$ Function Properties

The **Crypt$^{(e)}$** function in BM123-64 cipher consists of several $F^{V/e}_{n/m}$ functions having appropriate differential properties to construct the high probability DCs.

3.1.1. Differential Properties of $F_{2/2}$ Function

x_1 and x_2 are assumed as input parameters, with a pair (v, z) controlled vector of $F_{2/2}$ function. Therefore, the controlled element $F_{2/2}$ can be described as $F_{2/2}$ $(x_1, x_2, [v, z])$. Based on different distribution applied to different descriptions of $F_{2/2}$ in BM123-64, we have differential properties for each case as the following:

Case 1:
$$y_1 = vzx_1 \oplus vz \oplus vx_1 \oplus vx_2 \oplus v \oplus z \oplus x_1 \oplus 1$$
$$y_2 = vzx_2 \oplus vz \oplus vx_1 \oplus vx_2 \oplus zx_1 \oplus x_2 \oplus v \oplus z \oplus 1$$
$$\Pr[F_{2/2}(x_1, x_2, [v, z]) \oplus F_{2/2}(x_1 \oplus 1, x_2, [v, z]) = (1, 0)] = 2^{-2}.$$

Case 2:
$$y_1 = vzx_1 \oplus vzx_2 \oplus vx_1 \oplus vx_2 \oplus zx_1 \oplus zx_2 \oplus z \oplus v \oplus x_2$$
$$y_2 = vzx_1 \oplus vzx_2 \oplus vz \oplus vx_1 \oplus vx_2 \oplus zx_1 \oplus zx_2 \oplus x_1$$
$$\Pr[F_{2/2}(x_1, x_2, [v, z]) \oplus F_{2/2}(x_1 \oplus 1, x_2, [v, z]) = (1, 0)] = 2^{-1}.$$

Case 3:
$$y_1 = vx_2 \oplus x_2 \oplus x_1 \oplus v \oplus 1;$$
$$y_2 = vx_1 \oplus x_2$$
$$\Pr[F_{2/2}(x_1, x_2, [v, z]) \oplus F_{2/2}(x_1 \oplus 1, x_2, [v, z]) = (1, 0)] = 2^{-1}.$$

For each case of $F_{2/2}$ description, in order to get the (1, 0) output difference with the $(x_1 \oplus 1, 0)$ input difference and the (0, 0) controlled vector difference, the probability will be 2^{-2}, 2^{-1}, and 2^{-1} for **Case 1**, **Case 2**, and **Case 3**, respectively.

3.1.2. Differential Properties of $F_{16/64}$ and $F^{-1}{}_{16/64}$ Functions

In the same way, we can get the properties of SDDO-based functions $F_{16/64}$ and $F^{-1}{}_{16/64}$ using the properties above. Here, X is the input parameters for $F_{16/64}$ and $F^{-1}{}_{16/64}$ and (V, Z) is the controlled vector. Based on the properties of $F_{2/2}$, we can get the following:

Case 1: $\Pr[F_{16/64}(X, V, Z) \oplus F_{16/64}(X \oplus e_{16}, V, Z) = e_{16}] = 2^{-8}$
$\Pr[F^{-1}{}_{16/64}(X, V, Z) \oplus F^{-1}{}_{16/64}(X \oplus e_{16}, V, Z) = e_{16}] = 2^{-8}$

Case 2: $\Pr[F_{16/64}(X, V, Z) \oplus F_{16/64}(X \oplus e_{16}, V, Z) = e_{16}] = 2^{-4}$
$\Pr[F^{-1}{}_{16/64}(X, V, Z) \oplus F^{-1}{}_{16/64}(X \oplus e_{16}, V, Z) = e_{16}] = 2^{-4}$

Case 3: $\Pr[F_{16/64}(X, V, Z) \oplus F_{16/64}(X \oplus e_{16}, V, Z) = e_{16}] = 2^{-4}$
$\Pr[F^{-1}{}_{16/64}(X, V, Z) \oplus F^{-1}{}_{16/64}(X \oplus e_{16}, V, Z) = e_{16}] = 2^{-4}$

With the $F_{16/64}$ and $F^{-1}{}_{16/64}$ function designs, we can see that there is a total of four active layers $F_{2/2}$ within each construction, as shown in Figure 2.

3.2. Related-Key Boomerang of BM123-64

Here, we describe the way to generate related-key differential boomerangs on a full eight rounds of BM123-64 using the obtained properties in Section 3.1 with the key schedule generator.

We construct two related-key differentials as follows:

The first four rounds of related-key differential $\boldsymbol{E^0} = (\alpha \rightarrow \beta)$ holds with chosen input difference $\alpha = (0, 0)$ and key difference $(0, e_{32})$ from the first round to the fourth round with output difference $\beta = (0, 0)$. The key difference, based on the key schedule generator given in Table 1, is to control differential propagation and get high probability. The first related-key differential achieves this with probability $p = 2^{-16}, 2^{-8}$, and 2^{-8} for each case of $F_{2/2}$ description.

The second related-key differential $\boldsymbol{E^1} = (\gamma \rightarrow \delta) = (0, e_{32})$ " $(0, e_{16})$ covering the last four rounds with input difference $\alpha = (0, e_{32})$ and key difference $(0, e_{32})$ from the fifth round gives output difference $\delta = (0, e_{16})$ after the final transformation. The second related-key differential obtains probability $q = 2^{-16}, 2^{-8}$, and 2^{-8}, respectively. In total, the related-key differential characteristics constructed on a full eight rounds of BM123-64 designs give us the probability of 2^{-32} for **Case 1**, 2^{-16} for **Case 2**, and 2^{-16} for **Case 3**. Figure 2 illustrates the DCs propagation of some specific BM123-64 rounds in **Case 2** and **Case 3**.

The differential amplified boomerang can be explored based on the two related-key differentials above.

We can take any chosen plaintexts, namely (P, P^*, P', P'^*) with input difference $\alpha = (0, 0)$ described as $\Delta P = P \oplus P^* = P' \oplus P'^*$, to give respective ciphertexts with output difference $\beta = (0, 0)$ using related-keys (K, K^*, K', K'^*) defined as $\Delta K = K \oplus K^* = K' \oplus K'^* = (0, e_{16}, e_{32}, 0, 0, 0, 0, 0)$. We can hold the first four rounds related-key boomerang with probability of 2^{-16} for **Case 1**, 2^{-8} for **Case 2**, and 2^{-8} for **Case 3**. Furthermore, by pretending an additional round with intermediate differential values (I, I^*, I', I'^*) described as $I \oplus I^* = I' \oplus I'^* = (0, e_{32})$ using another related-key differences $\Delta K' = K \oplus K' = K^* \oplus K'^* = (0, e_{32}, 0, 0, 0, 0, 0, 0)$, we can build the second related-key boomerang on the last four rounds of 2^{-16} probability for **Case 1**, and 2^{-8} for both **Case 2** and **Case 3**.

In Table 2, the detailed DCs of a full eight rounds of BM123-64 with corresponding probabilities is shown. And, Figure 4 illustrates the DCs in $\mathbf{Crypt}^{(e)}$ function of BM123-64 with **Case 2** and **Case 3** at several rounds.

Table 2. Related-key DCs of a full eight rounds of BM123-64 in distinctive designs.

Round (r)	ΔX_r	$(\Delta U_r, \Delta Q_r)$	Probability		
			Case 1	Case 2	Case 3
1	$\alpha = (0, 0)$	$(0, e_{32})$	2^{-16}	2^{-8}	2^{-8}
2	$(e_{16}, 0)$	$(e_{16}, 0)$	1	1	1
3	$(0, 0)$	$(0, 0)$	1	1	1
4	$(0, 0)$	$(0, 0)$	1	1	1
Output	$\beta = (0, 0)$		2^{-16}	2^{-8}	2^{-8}
5	$(0, e_{32}) = \gamma$	$(0, e_{32})$	1	1	1
6	$(0, 0)$	$(0, 0)$	1	1	1
7	$(0, 0)$	$(0, 0)$	1	1	1
8	$(0, 0)$	$(0, e_{16})$	2^{-16}	2^{-8}	2^{-8}
FT	$(0, e_{16})$	$(0, 0)$	1	1	1
Output (ΔY)	$\delta = (0, e_{16})$				
Total			2^{-32}	2^{-16}	2^{-16}

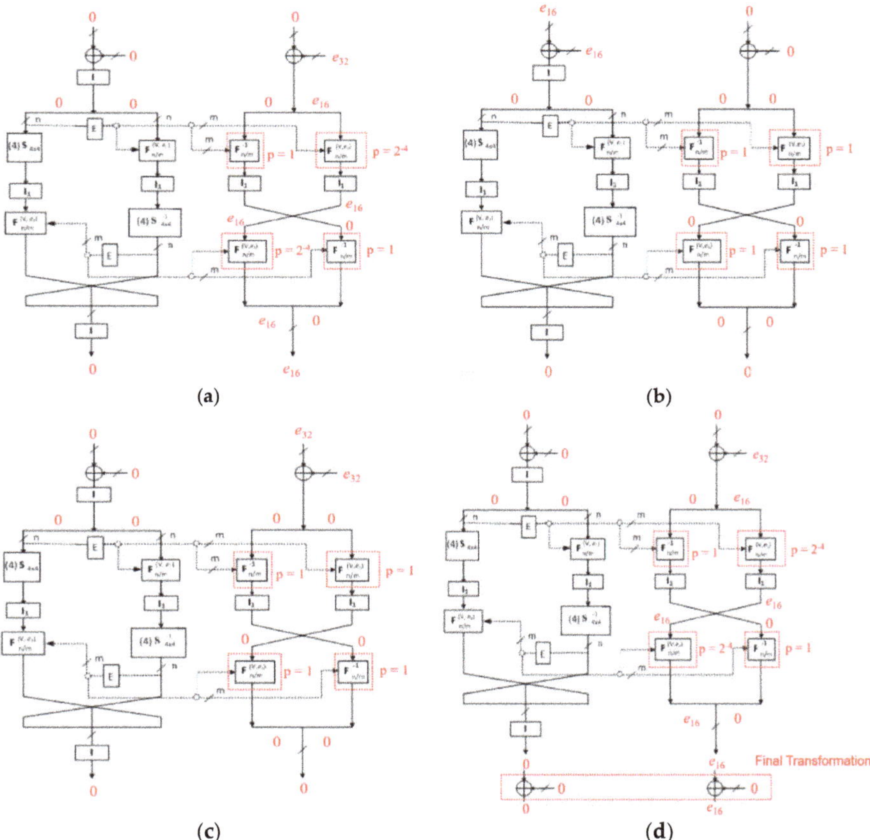

Figure 4. Differential characteristics (DCs) in **Crypt**$^{(e)}$ function at (**a**) the first round; (**b**) the second round; (**c**) the fifth round; and (**d**) the eighth round and final transformation of BM123-64 with **Case 2** and **Case 3**.

See (Appendix A) for more descriptions of DCs with **Case 1**.

3.3. Related-Key Amplified Boomerang Attack on the BM123-64 Designs

With the two obtained differentials, we expect that having $m^2 \cdot 2^{-128}$ of encrypted quartets for **Case 1** is right. For **Case 2** and **Case 3**, the expected number is $m^2 \cdot 2^{-96}$ right quartets. Furthermore, in the case of an ideal cipher, the related-key boomerang differentials applied on a full eight rounds of BM123-64 with probabilities of 2^{-128}, 2^{-96}, and 2^{-96} ($2^{-64} \cdot p^2 \cdot q^2$) for each case, respectively. For the least expected right quartets number of 8, we take a set of 2^{66} pairs of plaintexts ($m^2 \cdot 2^{-128} = 2^3$) for the attack process in **Case 1** and 2^{50} pairs of plaintexts ($m^2 \cdot 2^{-96} = 2^3$) for the attacks in **Case 2** and **Case 3**.

The attacker follows the below steps for a full related-key attack method on BM123-64:

(1) We pick a set of 2^{66} pairs of plaintexts (P_j, P_j *), (j = 1, ... , 2^{66}), then we expand into another set of 2^{131} quartets of plaintexts, denoted as (P_i, P_i *, P_i ′, P_i ′*), (i = 1, ... , 2^{131}) in **Case 1**, or 2^{50} pairs of plaintexts (P_j, P_j *), (j = 1, ... , 2^{50}) and generate 2^{99} quartets of plaintexts (P_i, P_i *, P_i ′, P_i ′*), (i = 1, ... , 2^{99}) in **Case 2** and **Case 3**, with input difference $\alpha = (0, 0)$. We ask for encryption of all the quartets (P_i, P_i *, P_i ′, P_i ′*) using the related-keys (K, K^*, K', K'^*) difference, described as two terms of relation: $\Delta K = K \oplus K^* = K' \oplus K'^* = (0, e_{16}, e_{32}, 0, 0, 0, 0, 0)$ and $\Delta K' = K \oplus K' = K^* \oplus K'^* = (0, e_{32}, 0, 0, 0, 0, 0, 0)$ to output respective quartets of ciphertexts (C_i, C_i *, C_i ′, C_i ′*).

(2) We do XOR with all possible values of C_i and C_i', C_i *, and $C_i'^*$ for each i value, then check whether the output result is $(0, e_{16})$ and store all these difference values to apply in the previous eight rounds.

(3) By this way, at the final transformation, we expect to hold a 64-bit subkey including K_1 and K_3, then get the remaining subkeys (K_1^*, K_3^*), (K_1', K_3'), and ($K_1'^*$, $K_3'^*$) of the quartets of subkeys.

    (a) Similarly, at the eighth round, we ask for decryption of all quartets of ciphertexts values obtained from Step 2 with subkey quartets of K_1 and K_3 to hold 64-bit input values (X_j, X_j^*, X_j', $X_j'^*$) at the left side process of round function.

    (b) We do XOR with all possible values of X_j and X_j', X_j *, and $X_j'^*$ for each j value, then check whether the output result is 0.

(4) After passing Step 3, all values of quartets of two subkeys K_1 and K_3 are explored. We can do brute force attacks to obtain the remaining 192-bit subkeys (K_2, K_4, K_5, K_6, K_7, K_8) with all K_1 and K_3.

4. Results and Discussion

With **Case 1**, the proposed attack requires 2^{66} pairs of plaintexts and 2^{67} related-key chosen plaintexts in data complexity, since the related-key DC is 2^{-32}. Furthermore, it needs about 2^{70} ($=2^{67} \times 8$) bytes of memory.

With **Case 2** and **Case 3**, the attack requires 2^{50} pairs of plaintexts and 2^{51} related-key chosen plaintexts in data complexity, while the related-key DCs are 2^{-16}. The attack takes about 2^{54} ($=2^{51} \times 8$) bytes of memory.

At Step 1 of the attack, the complexity of time is a unit of 2^{67} encryptions (**Case 1**), or 2^{51} encryptions (**Case 2** and **Case 3**) of the full eight rounds of BM123-64. Each quartet of ciphertext is planned to achieve Step 2 of the attack with 2^{-64} probability. In addition, we will get 2^{67} ($=2^{131} \times 2^{-64}$) right quartets of ciphertext (**Case 1**) or 2^{35} right quartets of ciphertext ($=2^{99} \times 2^{-64}$) (**Case 2** and **Case 3**) that will achieve Step 2. At Step 3 and Step 4, the complexity of time is a unit of 2^{62} ($=2^{64} \times 4 \times 1/8 \times 1/2$) and 2^{65} ($=2^{64} \times 1 \times 2$) for a full eight rounds of BM123-64 encryptions, respectively. Finally, the results show that all the attacks require total time complexity of 2^{67} ($\approx 2^{67} + 2^{62} + 2^{65}$) (**Case 1**) or 2^{65} ($\approx 2^{51} + 2^{62} + 2^{65}$) (**Case 2** and **Case 3**) for a full eight rounds of BM123-64 encryptions on average. In Table 3, a comparison of cryptanalysis results between the scheme proposed and other data-dependent ciphers in terms of complexity of data and time is given.

At Step 4 of the attack, outputting a wrong quartet of subkey takes 2^{-64} of probability. With our cryptanalysis methods, the results shown with the output possibility in case of a wrong key is lower than the ideal case. These proposed related-key amplified boomerang attacks can potentially exploit the BM123-64 constructions at all three specific cases.

Table 3. Cryptanalysis results on constructions based on DDP (Data-Dependent Permutation), DDO (Data-Dependent Operation), and SDDO (switchable data-dependent operation).

Block Cipher	Total Rounds	Complexity Data/Time	Key Bits Recovery
DDP-64	10/10	2^{54} RCP/2^{54}	22
CHESS-64	8/8	2^{44} RCP/2^{44} 2^{39} RCP/2^{39} 2^{44} RCP/2^{108} 2^{39} RCP/2^{122}	20 6 128 128
DDO-64V$_1$	8/8	$2^{35.5}$ RCP/$2^{65.5}$	
DDO-64V$_2$	8/8	2^{3} RCP/2^{31}	
MD-64	8/8	$2^{43.1}$ RCP/2^{95}	
BMD-128	7/8	2^{79} RCP/2^{129}	
KT-64	8/8	$2^{45.5}$ RCP/$2^{65.17}$	
XO-64	8/8	2^{44} RCP/2^{65}	
BM123-64 (**Case#1**) (*)	8/8	2^{67} RCP/2^{67}	
BM123-64 (**Case#2**) (*) BM123-64 (**Case#3**) (*)	8/8	2^{51} RCP/2^{65}	

(*) our proposed attack results; RCP denotes the Related-key Chosen Plaintext.

5. Conclusions

This paper proposes effective attacks on one of the recent SDDO-based constructions, BM123-64. The methods addressed in the study present main issues in security mechanisms of ciphers based on data-dependent operations that is used in many cipher designs for high speed transformation and lightweight targets. The simple key schedule generator with basic parameters when changed and substituted leads to a possibility of exploiting the structure weaknesses by related-key cryptanalysis. The work presented related-key amplified boomerang attacks on a full eight rounds of BM123-64 in distinctive designs with effective complexity results. This shows that with **Case 1**, it requires 2^{66} related-key chosen plaintexts and 2^{67} encryptions consumptions, and 2^{50} related-key chosen plaintexts and 2^{65} encryptions consumptions are required with **Case 2** and **Case 3**. The results of this study can be applied on many construction designs of these types of ciphers. Along with some new cryptanalysis techniques like Fr Trust or RARE, our research will further enhance performance and is expected to develop novel approaches for a wide range of applications and devices in the IoT environment.

Author Contributions: Conceptualization, T.S.D.P., C.L.; Methodology, T.S.D.P.; Writing-Original Draft Preparation, T.S.D.P.; Writing-Review & Editing, T.S.D.P.; Supervision, C.L.; Funding Acquisition, C.L.

Funding: This work was supported by the Korea Institute of Energy Technology Evaluation and Planning (KETEP) and the Ministry of Trade, Industry & Energy (MOTIE) of the Republic of Korea (Development of Cyber Security Technology for Non-safety Grade Control System (DCS) in Nuclear Power Plants, No. 20161510101810).

Conflicts of Interest: The authors declare no conflict of interest.

Appendix A

Figure A1 illustrates the differential propagation at several rounds of BM123-64 cipher with **Crypt**$^{(e)}$ function in **Case 1**.

We pick a set of 2^{66} pairs of plaintexts (P_j, P_j *) and expand to 2^{131} quartets of plaintexts (P_i, P_i*, P_i', P_i'*). The SDDO-based $F^{V/e}_{n/m}$ function in BM123-64 structure includes $F_{16/64}$ and $F^{-1}_{16/64}$, which has four layers with eight $F_{2/2}$ element function in each. Based on the differential properties mentioned in Section 3.1, we can construct two related-key boomerangs on a full eight rounds of BM123-64 in Case 1 with total probability of 2^{-32}. The first four rounds of related-key boomerang holds with the probability $p = 2^{-16}$; the last four rounds of related-key boomerang also obtains with the probability

$q = 2^{-16}$. It requires about 2^{67} chosen plaintexts in complexity of data as input, and the memory required is about 2^{70} ($=2^{67} \times 8$) bytes.

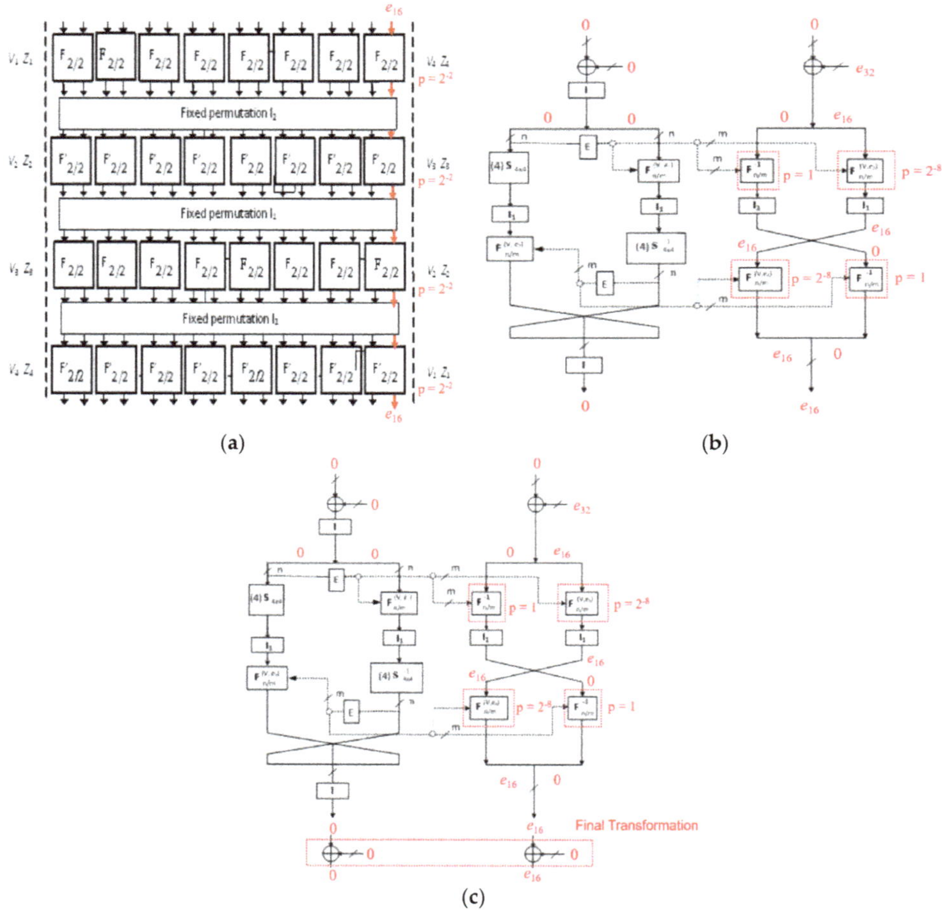

Figure A1. (**a**) The differential propagation of $F_{16/64}$, $F^{-1}{}_{16/64}$ function and the DCs in **Crypt**$^{(e)}$ function at (**b**) the 1st round and (**c**) the eighth round and final transformation of BM123-64 with **Case 1**.

References

1. Bac, D.; Minh, N. High-Speed Block Cipher Algorithm Based on Hybrid Method. In *Ubiquitous Information Technologies Applications*; Lecture Notes in Electrical Engineering; Springer: Berlin/Heidelberg, Germany, 2014; Volume 280, pp. 285–291.
2. Moldovyan, N. On Cipher Design Based on Switchable Controlled Operations. In *MMM-ACNS, LNCS*; Springer: Berlin/Heidelberg, Germany, 2003; Volume 2776, pp. 316–327.
3. Bac, D.; Minh, N.; Duy, H. An Effective and Secure Cipher Based on SDDO. *Int. J. Comput. Netw. Inf. Secur.* **2012**, *4*, 1.
4. Bac, D.; Minh, N.; Duy, H. New SDDO-Based Block Cipher for Wireless Sensor Network Security. *Int. J. Comput. Netw. Inf. Secur.* **2010**, *10*, 54–60.

5. Minh, N.; Luan, N.; Dung, L. KT-64: A New Block Cipher Suitable to Efficient FPGA Implementation. *IJCSNS Int. J. Comput. Netw. Inf. Secur.* **2010**, *19*, 10–18.
6. Minh, N.; Duy, H.; Dung, L. Design and Estimate of a New Fast Block Cipher for Wireless Communication Devices. In Proceedings of the International Conference on Advanced Technologies for Communications, Hanoi, Vietnam, 6–9 October 2008; pp. 409–412.
7. Moldovyan, N.; Moldovyan, A.; Sklavos. Controlled Elements for Designing Ciphers Suitable to Efficient VLSI Implementation. *Telecommun. Syst. J.* **2006**, *32*, 149–163. [CrossRef]
8. Kang, J.; Jeong, K.; Lee, C.; Hong, S. Distinguishing attack on SDDO-based block cipher BMD-128. In *Ubiquitous Information Technologies and Applications*; Springer: Berlin/Heidelberg, Germany, 2014; Volume 280, pp. 595–602.
9. Phuc, T.S.D.; Lee, C.; Xiong, N. Cryptanalysis of the XO-64 Suitable for Wireless Systems. *Wirel. Pers. Commun.* **2017**, *93*, 589–600. [CrossRef]
10. Izotov, B.V.; Moldovyan, N.; Moldovyan, A. Controlled Operations as a Cryptographic Primitive. In *Information Assurance in Computer Networks*; Springer: Berlin/Heidelberg, Germany, 2001; Volume 2052, pp. 230–241.
11. Kang, J.; Jeong, K.; Yeo, S.; Lee, C. Related-key Attack on the MD-64 Block Cipher Suitable for Pervasive Computing Environment. In Proceedings of the International Conference on Advance Information Networking and Application Workshops, Fukuoka, Japan, 26–29 March 2012; pp. 726–731. [CrossRef]
12. Lee, C.; Kim, J.; Sung, J.; Hong, S.; Lee, S. Security analysis of the full-round DDO-64 block cipher. *J. Syst. Softw.* **2008**, *84*, 2328–2335. [CrossRef]
13. Moldovyan, N.; Moldovyan, A. Data-driven Ciphers for Fast Telecommunication Systems. In *Auerbach Publication*; Talor & Francis Group: New York, NY, USA; London, UK, 2008; pp. 77–185, ISBN 1420054112 9781420054118.
14. Biham, E.; Dunkelman, O.; Keller, N. Related-key boomerang and rectangle attacks. In *Advances in Cryptology—EUROCRYPT'05, LNCS*; Springer: Berlin/Heidelberg, Germany, 2005; Volume 3494, pp. 507–525.
15. Kelsey, J.; Kohno, T.; Schneier, B. Amplified Boomerang Attacks against Reduced-Round MARS and Serpent. In *Proceedings of Fast Software Encryption 7*; Lecture Notes in Computer Science; Springer: Berlin/Heidelberg, Germany, 2000; Volume 1978, pp. 75–93. [CrossRef]
16. Wagner, D. The Boomerang Attack. In *Proceedings of Fast Software Encryption 6*; Lecture Notes in Computer Science; Springer: Berlin/Heidelberg, Germany, 1999; Volume 1636, pp. 156–170. [CrossRef]

Article

Ephemeral ID Beacon-Based Improved Indoor Positioning System

Jinsu Kang, Jeonghoon Seo and Yoojae Won *

Department of Computer Science and engineering, Chungnam National University, Daejeon 34134, Korea; kangjs1798@cnu.ac.kr (J.K.); sjh9309@cnu.ac.kr (J.S.)
* Correspondence: yjwon@cnu.ac.kr; Tel.: +82-42-821-6294

Received: 28 September 2018; Accepted: 8 November 2018; Published: 10 November 2018

Abstract: Recently, the rapid development of mobile devices and communication technologies has dramatically increased the demand for location-based services that provide users with location-oriented information and services. User location in outdoor spaces is measured with high accuracy using GPS. However, because the indoor reception of GPS signals is not smooth, this solution is not viable in indoor spaces. Many on-going studies are exploring new approaches for indoor location measurement. One popular technique involves using the received signal strength indicator (RSSI) values from the Bluetooth Low Energy (BLE) beacons to measure the distance between a mobile device and the beacons and then determining the position of the user in an indoor space by applying a positioning algorithm such as the trilateration method. However, it remains difficult to obtain accurate data because RSSI values are unstable owing to the influence of elements in the surrounding environment such as weather, humidity, physical barriers, and interference from other signals. In this paper, we propose an indoor location tracking system that improves performance by correcting unstable RSSI signals received from BLE beacons. We apply a filter algorithm based on the average filter and the Kalman filter to reduce the error range of results calculated using the RSSI values.

Keywords: Internet of things (IoT); location-based service (LBS); indoor localization; beacon; received signal strength indication (RSSI); Kalman filter; average filter

1. Introduction

With the development of information and communication technology due to the Fourth Industrial Revolution, the time and space restrictions on access to information are disappearing. Any device with Internet access grant the user access any desired information, regardless of time or place. Recently, technologies are being developed that predict and provide information that the user needs in accordance with the current situation or condition of the user before user finds information. As part of this research, location-based services (LBS) have emerged that provide information based on user's location.

An LBS collects user location information through a communication network or Internet of things (IoT) equipment, and provides users with specific information and services such as traffic, weather, and nearby landmarks or events, based on the collected location information. Due to factors such as the evolution of smartphones, the proliferation of high-speed telecommunications network infrastructure, real-time location tracking services, and increased demand for related products and services such as location-based advertising and marketing for customers, the LBS market is currently experiencing rapid growth [1,2]. Recently, the importance of indoor space has been emphasized as most activities such as business, entertainment, and shopping are performed in the indoor space. As a result, various types of location-based services provided for indoor space have also started to attract attention. The core technology of LBS is localization technology that tracks the user's location. There are various kinds of positioning technologies. Among these, Global Navigation Satellite System (GNSS) such as GPS is generally used as positioning technology of LBS [3].

Although GPS shows good performance in the outdoor space, it is difficult to trace the user's location in the indoor space using the GPS because the reception of signals is not smooth in the indoor spaces. Therefore, various indoor positioning technologies to replace GPS are currently being researched and developed. One prevalent approach to indoor positioning technology is to install devices that transmit signals using Bluetooth, sound waves, etc., in an indoor space, and then calculate user position based on the signals received by their wireless and mobile devices such as Wi-Fi receivers and smart phones. In contrast, a different method measures user position using hardware that is built into users' devices. The former approach is capable of receiving more stable position information, but comes with costs of additional equipment and the installation process. In the latter method, although no additional equipment is required and financial cost is reduced, the positioning stability and accuracy are low. Furthermore, the hardware can quickly consume the user's device battery owing to the ongoing calculation processes. Among the indoor positioning techniques, there are methods using BLE (Bluetooth Low Energy) beacons. To measure a user's position in an indoor space using a beacon, a plurality of beacons is installed in the area, and the distance between each beacon and a mobile device is measured based on the received signal strength indicator (RSSI) of the beacon. This method locates the user by applying a positioning algorithm, such as the triangulation method, under the assumption that the calculated distance between each beacon and mobile device is accurate. However, it is still difficult to accurately determine indoor position using RSSI because the signal is unstable owing to obstacles in measurement space, the surrounding environment, weather, humidity, and other signals. Thus, to achieve accurate indoor positioning, it is necessary to compensate for unstable RSSI. In this paper, we propose an improved indoor positioning system that measures user's position in indoor space. For this purpose, we applied the appropriate filter algorithm based on the extended Kalman filter (EKF) to the unstable RSSI value to reduce the error range of the computed distance.

The contributions of this work are as follow.

1. We reduced the error range of RSSI by applying filter algorithm based on EKF and average filter.
2. We construct a system with enhanced security by applying ephemeral ID technology when identifying beacon.

This paper is organized in six sections. Section 2 reviews the related work and Section 3 describes the indoor positioning method using beacons. Section 4 presents the indoor positioning system, Section 5 shows the application of the filter algorithm and finally, the conclusions and future lines of work are presented in Section 6.

2. Related Work

2.1. Performance Improvement of Indoor Positioning System

Due to the increasing demand of indoor LBS, many researches on indoor localization technology have been conducted. Many of the proposed approaches utilize beacons for indoor location tracking.

Inoue et al. [4] proposed a positioning system for an indoor pedestrian navigation service that used mobile phones and BLE beacons in a server-free environment. Multiple beacons were installed at various points in the room and mobile devices collected RSSI broadcast by the beacons. After receiving the signal data from a beacon, the relative position of the user was estimated based on analysis that used the probability inference algorithm.

Lin et al. [5] proposed a mobile-based indoor location system that uses mobile applications as an iBeacon solution based on BLE technology to improve the efficiency of the emergency room. They collected the RSSI of the surrounding beacons and calibrated the RSSI by applying an average filter. Then, they repeatedly applied the average filter to RSSI, which was repeated many times, and found the nearest beacon and used it for the calculation.

Rida et al. [6] proposed an indoor positioning system based on the RSSIs of BLE technology. They installed a beacon on the ceiling of the indoor space and calculated the location of the user in the indoor space by applying a dilatation algorithm that is easy for any hardware to implement because of the algorithm's low complexity.

Momose et al. [7] proposed an indoor positioning system that improved the accuracy of beacon distance measurement by applying a particle filter based on the space's floor plan. By using the proximity of BLE beacons as well as acceleration and geomagnetic sensors to design algorithms.

Pelant et al. [8] created a Ray-Launching based simulation model to study the performance of BLE technology in indoor localization, where RSS fingerprinting map was created based on theoretical and measurements results; their results show that accuracy highly depends on the number of sectors in the environment and on the number of beacons considered.

Luo et al. [9] proposed a method of arranging robots in an indoor space using Wi-Fi and radio-frequency identification (RFID) fingerprinting. They collected Wi-Fi and RFID signals at multiple points in the room. The fingerprint of Wi-Fi was initially used to place educational robots (Erobs) in large areas and then direct the robot to a location close to the target location. The RFID fingerprint would then be used to direct the Erob to the target location.

Dong et al. [10] proposed a solution combining feature clustering and a wireless sensor network to increase the efficiency of robot position recognition in indoor spaces. They also integrated robotic position recognition with wireless sensor networks to improve robustness.

2.2. Google's Eddystone

Eddystone is an open-source beacon format published by Google, and has open-source protocol specifications and development tools, making it easy to extend and operate on a variety of platforms.

Eddystone supports four frames. The first is the UID Frame. UID frame broadcasts a unique ID of the beacon. UID frame consists of 10 Byte namespace and instance of 6 Bytes. The namespace is used to distinguish groups of beacons, and the instance is used to identify each beacon in the same group [11].

The second is URL frame. URL frame is a frame that allows the user to directly use the existing contents in the physical web by advertising the URL. URL frame broadcasts a URL using a compressed encoding format in order to fit more within the limited advertisement packet. Once decoded, the URL can be used by any client with access to the internet [12].

The third is the TLM frame. TLM frame is a frame that transmits data that can monitor the beacon status, such as the beacon sensor, the battery, and the transmission status. TLM frame must be paired with a frame providing an identifier of UID or URL type because the packet does not contain a beacon identifier [13].

The last is the EID (ephemeral id) frame. The EID broadcasts a temporary identifier that changes periodically at a rate determined when it first registers with the web service. An EID can be resolved by a registered service, but appears to be randomly changed to other observers who are not registered. This frame type is used for security and privacy enhancement [14].

To generate an EID that can only be identified by registered users, three data are required. The first is identity key. The identity key is the value shared by beacon and the server resolving the ephemeral identity key. The second is rotation period exponent scalar. This value is a value that determines the period in which the temporary ID changes. This value has a value between 0 and 15. When this value is K, a new temporary ID is generated every 2^K s. The last is a 32-bit beacon-specific timestamp value.

Figure 1 shows the EID generation process. beacon and server each have a public key and a private key pair. First, when an EID registration request comes in, the server returns its public key, along with its supported rotation period exponents range. Then, beacon computes the ECDH shared secret by multiplying its secret key and server's public key. Then, the shared secret is encrypted with the HKDF-256 algorithm, and the upper 16 bytes are cut out to generate the identity key.

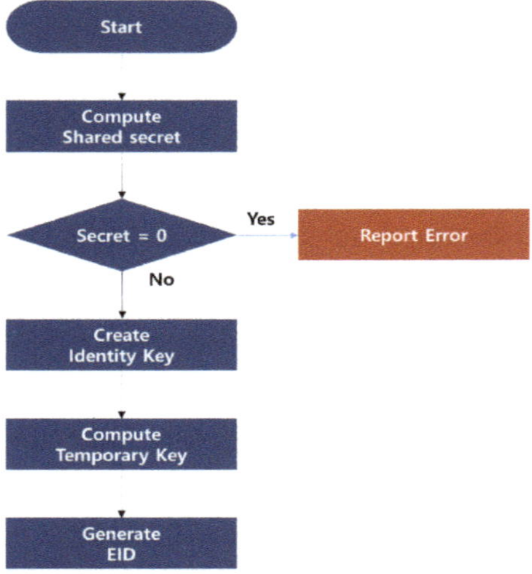

Figure 1. Eddystone-EID generate process.

Then, the process of generating the temporary key using the identity key and finally the process of generating the EID are performed. Temporary key is generated by encrypting the data configured as Table 1 with AES-128 algorithm using the identity key as the encryption key. Then, the same procedure is repeated for the data as shown in Table 2, using the temporary key as the encryption key. EID is completed by taking the upper 8 bytes of the encryption result [15].

Table 1. Structure of data for temporary key generation.

Byte Offset	Description
Byte 0–10	Padding (0 × 00)
Byte 11	Salt (0 × ff)
Byte 12–13	Padding (0 × 00)
Byte 14–15	Top 16 bit of time counter

Table 2. Structure of data for EID Generation.

Byte Offset	Description
Byte 0–10	Padding
Byte 11	Rotation period exponent
Byte 12–15	Time counter 32 bit

3. Indoor Positioning Method Using Beacon

This section will describe how to measure the position in a room using beacons. To measure indoor location using beacons, a positioning algorithm, such as the trilateration method, is used. This means that three or more beacons must be installed in the measurement space.

As shown in Figure 2, the interior space can be represented on a two-dimensional plane. To apply trilateration, the location of each beacon and mobile device is indicated using (x, y) coordinates. Let the position coordinates of each beacon be $(x1, y1)$, $(x2, y2)$, $(x3, y3)$ and the position of the mobile device be (a, b). At this time, the following set of equations is established:

$$\begin{array}{l}(\mathrm{x}1-\mathrm{a})^2+(y1-b)^2=d1^2\\(\mathrm{x}2-\mathrm{a})^2+(y2-b)^2=d2^2\\(\mathrm{x}3-\mathrm{a})^2+(y3-b)^2=d3^2.\end{array} \quad (1)$$

If all the values of the beacon position coordinates, $(x1, y1)$, $(x2, y2)$, $(x3, y3)$, and the distances $d1$, $d2$, and $d3$ between the mobile device and the beacon are known, then, the mobile device position (a, b) can be obtained. Because the location coordinates of the beacon are known in advance and the distance between the beacon and the mobile device can be measured using the RSSI, the position of the mobile device in an indoor space can be determined and indicated on a two-dimensional plane [16].

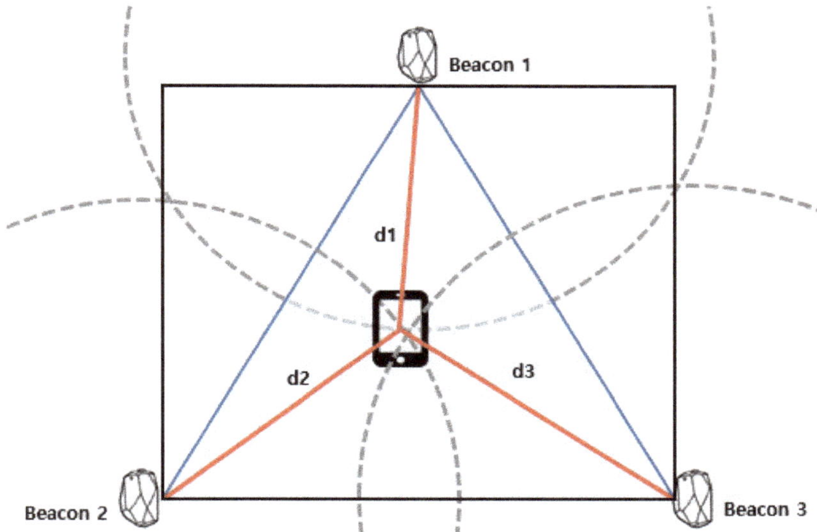

Figure 2. Indoor location measurement using trilateration.

The values of $d1$, $d2$ and $d3$, which are the distance between the beacons and the mobile device, must be accurate to determine the correct position of the mobile device in the indoor space using the trilateration method. The formula for calculating the distance between a beacon and the terminal using the RSSI is as follows:

$$d = 10^{\frac{Txpower - \mathrm{RSSI}}{10*N}}, \quad (2)$$

where Tx power is the intensity at which a beacon transmits a signal and N is a correction constant with a value associated with propagation loss. Because the Tx power and N value are constant, the value of the distance d calculated according to the RSSI changes. The RSSI is irregularly measured because the RSSI is affected by the environment of the measurement space. If a distance value is incorrect because of unstable RSSI, the scenarios depicted in Figures 3–5 occur.

In the event of an unstable RSSI, the circles of reception generated by the RSSI signals from each beacon do not meet. Without circle contact, the position of the mobile device cannot be calculated [17].

Figure 4 depicts the situation where only two of the three circles generated by the RSSI measured from each beacon are encountered. In this case, because two points of contact are generated in the circle, two locations of mobile devices can be calculated. However, the exact location of the mobile device cannot be obtained without the orientation provided by the third signal because it may appear that the mobile device is outside of the overlap when it actually may be inside the triangle [17].

Figure 5 illustrates the case where all three circles generated by the RSSI measured from each beacon overlap; however, they fail to meet at a single, specific point. The calculated distance differs from the actual distance, because although the three generated circles overlap, they are not seen at a single point but rather may be encountered at three points. In some cases, a circle may be encountered outside the triangle. Thus, the exact location of a mobile device cannot be determined.

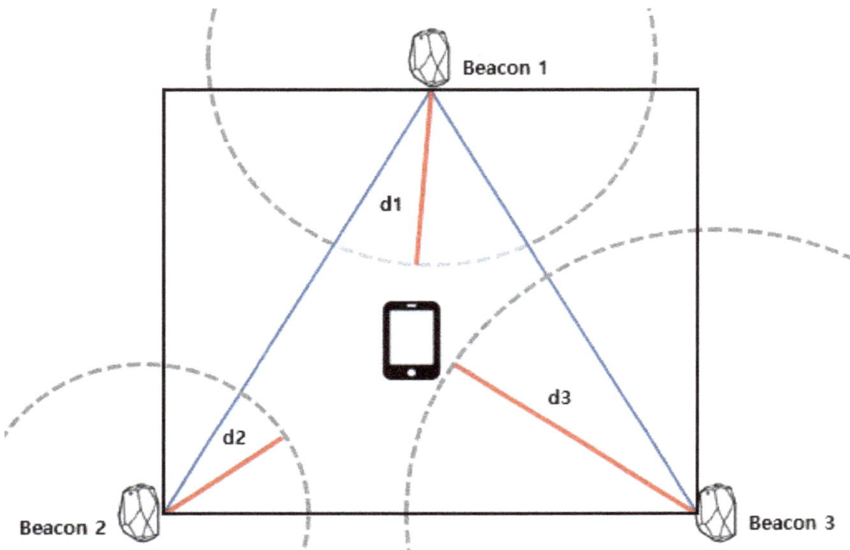

Figure 3. Three circles in trilateration do not overlap.

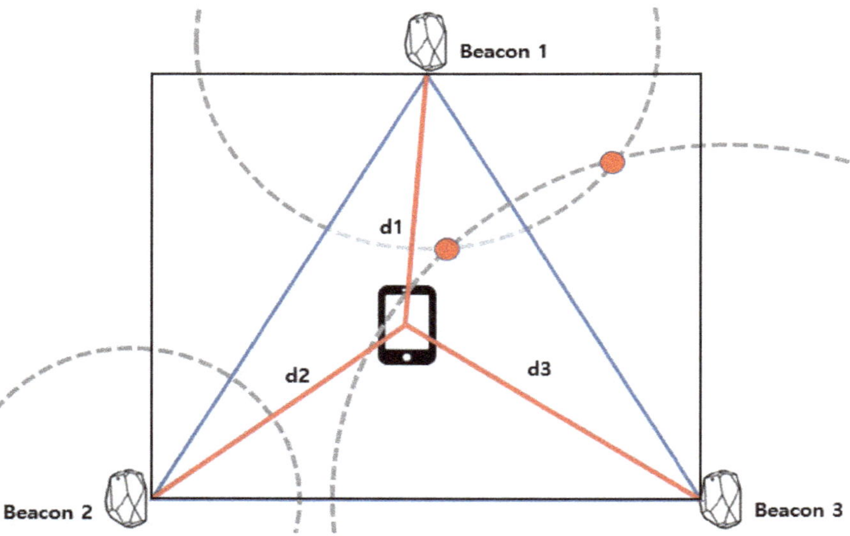

Figure 4. Two circles overlap between two intersecting points.

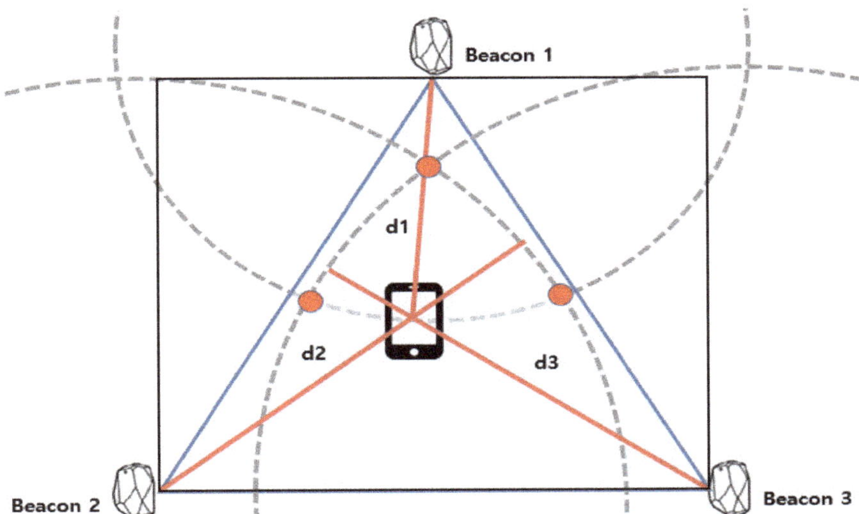

Figure 5. Three circles do not intersect at a single point as required for trilateration.

4. Proposed Indoor Localization System

The proposed indoor positioning system consists of three major components. First, a data acquisition module collects the beacon RSSI values received by mobile devices from those mobile devices. The data is then corrected by the filter algorithm and a data processing module applies a positioning algorithm to the filtered results. The third major component, the data management module, stores and manages data processing results in a database. Figure 6 illustrates the structure of the proposed system.

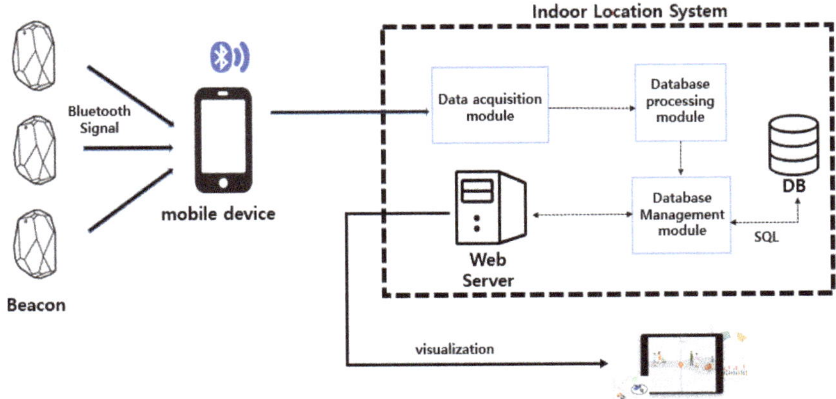

Figure 6. Configuration of proposed indoor positioning system.

4.1. Data Acquisition Module

The data collection module collects various data from the user's mobile device. The user's mobile device runs an application that collects data from a beacon using Bluetooth communication and sends it to the data acquisition module of the indoor positioning system. The data acquisition module communicates with the mobile device through the mobile application and receives the message authentication code (MAC) value of the device, the identifier of the beacon, the Tx power, and the RSSI. The identifier of the beacon is collected to identify the user's absolute location in the indoor space.

In this system, Eddystone EID frame is used as an identifier to distinguish beacons belonging to the system. The EID frame is composed of 8 bytes and the data configuration to transmit in actual communication is as shown in Table 3. The 0th byte is the area that defines the frame type and its EID is set to 0×30. The first byte is the Tx power. Next, the temporary ID for identifying the beacon is composed of 8 bytes [14].

Table 3. Eddystone-EID structure.

Byte Offset	Description
Byte 0	Frame Type (EID = 0×30)
Byte 1	Tx Power
Byte 2–9	8 Byte Ephemeral Identifier

4.2. Data Processing Module

The data processing module preprocesses the data to reduce the error range using a filtering algorithm to correct the RSSI data received from the data acquisition module. This module uses the RSSI data and Tx power data of the beacons collected by communicating with the mobile devices in the collection module to calculate the distance between each beacon and the mobile device, and then uses those distances to calculate the position of the user in the indoor space. The data calibration process used in this module is explained in Section 5.

4.3. Data Management Module

The data management module specifies the location of a user's mobile device and manages various data required by the indoor positioning system after the data processing module has analyzed the data collected by the acquisition module.

Information about the computed location is stored in the local database of the indoor positioning system. To manage the beacons in the system, beacon identifiers and information about the actual location where each beacon is located in the indoor space are also stored in the system database, as are the distance data measured from each beacon in the indoor space. Additionally, several tables for aggregation are defined and to further facilitate data management. In the data management module, the Eddystone Configuration GATT Service can be used to communicate with the beacons to set Tx power, signal interval, and so on [18].

5. Application and Measurement Results of Filter Algorithm

5.1. Apply Filter Algorithm

When an RSSI broadcast by a beacon near a mobile device is collected by an application running on the mobile device, the data processing module of the positioning system corrects the value by applying a filter algorithm. The application of the filter algorithm is divided into two steps. The first is the application of the average filter. As explained previously, because RSSI is highly influenced by ambient noise, new results are obtained for every measurement even if measurements have already been performed at the same position. To solve this problem, after collecting the RSSI several times, the average filter was applied to correct the RSSI value. The formula of the average filter is delineated in Equation (3).

$$\overline{x}_k = \frac{k-1}{k}\overline{x}_{k-1} + \frac{1}{k}x_k \quad (3)$$

In Equation (3), x represents data, that is, RSSI, and k indicates the measurement time. The value of the k-th data is calculated by weighting the measured value at the k-th measurement and the calculated values before that. The following is the application of the Kalman Filter. Kalman filter is applied to the corrected value once using the average filter, and the correction process is repeated. A Kalman filter is a recursive filter that tracks the state of a dynamic system that contains noise. The operation of the Kalman filter is shown Figure 7. [19] The application of the Kalman filter consists of a state prediction step and a two-part measurement update step. The data are corrected while continuously repeating this process.

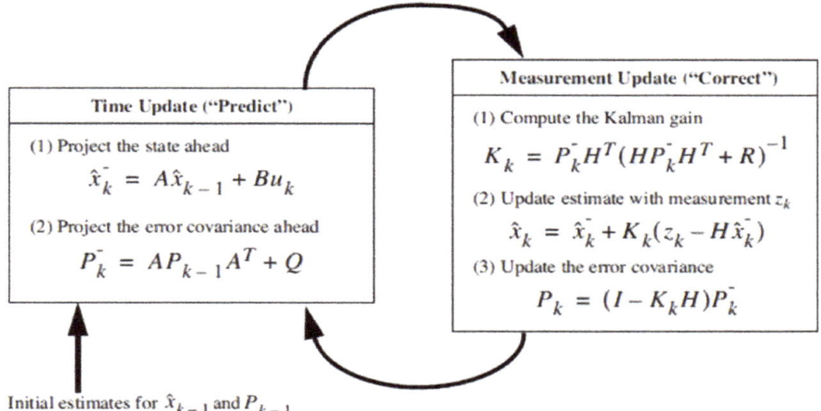

Figure 7. The process of tracking the state of a Kalman filter.

When new data are received, the state prediction step predicts the state of the new data based on the previously estimated state. The measurement update step calculates the current state of the data by determining which are more reliable based on the state of the data predicted in the state prediction step and actual measured data [20].

5.2. RSSI Correction Result

This section discussed the results of calibrating the RSSI by applying a filter algorithm. For the experiment, we installed a beacon in a specific space, received the RSSI on the mobile device, and then used it to calculate the distance between the beacon and the mobile device. To reduce the influence of other factors on the comparison of the result, other factors such as the *Tx* power of the beacon and the advertise interval were set equal to a predetermined value before performing the measurement. The performance was analyzed by comparing measurement results obtained without applying the filter algorithm with those obtained using the filter algorithm.

Figure 8 shows the results of RSSI measurements taken at a distance of 2 m from a single beacon for several seconds. Figure 8a shows the raw data, and Figure 8b shows the corrected data. As observed in the graphs, irregular data were recorded for each measurement even if the RSSI data without the algorithm had been measured at the same position. Conversely, it was confirmed that the data corrected by applying the filter algorithm were stable and constant compared to the raw data.

Figures 9–11 show the results of distance measurements at various distances. The distances were measured using RSSI at distances of 4 m, 7 m and 11 m, respectively, and then plotted as a graph. Compared with the raw data, the corrected data has a more stable and has an error value within 10%. However, it has been confirmed that the measurement distance is less accurate at a distance of 11 m or more.

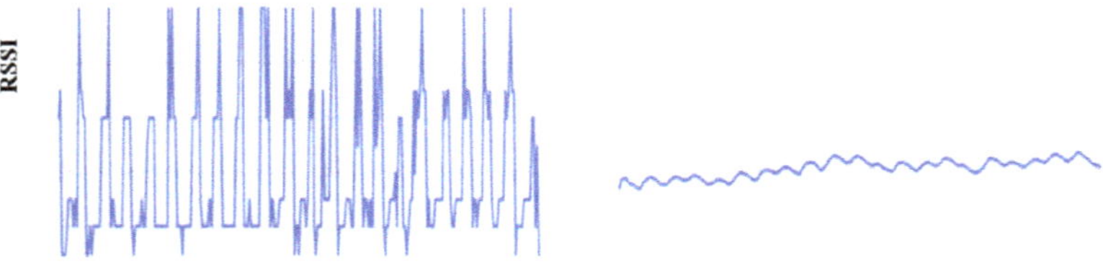

Figure 8. RSSI measurement results: (a) Raw data; (b) Corrected data.

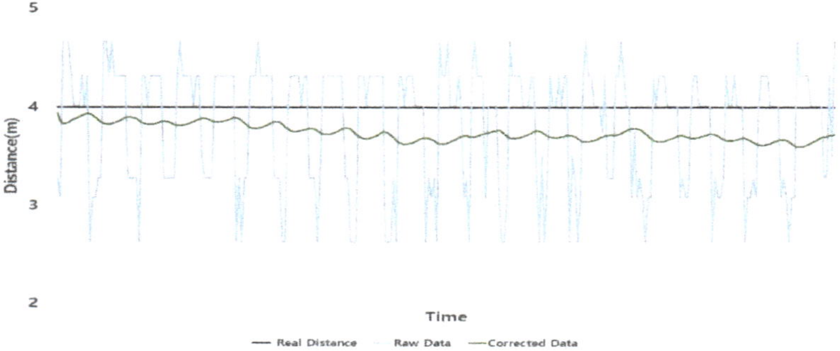

Figure 9. Distance measurement result (4 m).

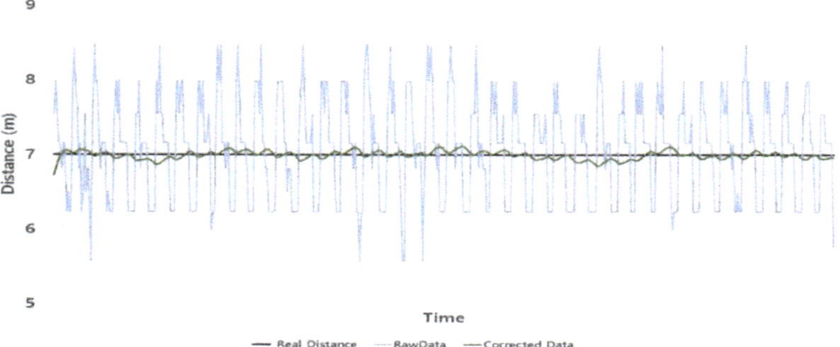

Figure 10. Distance measurement result (7 m).

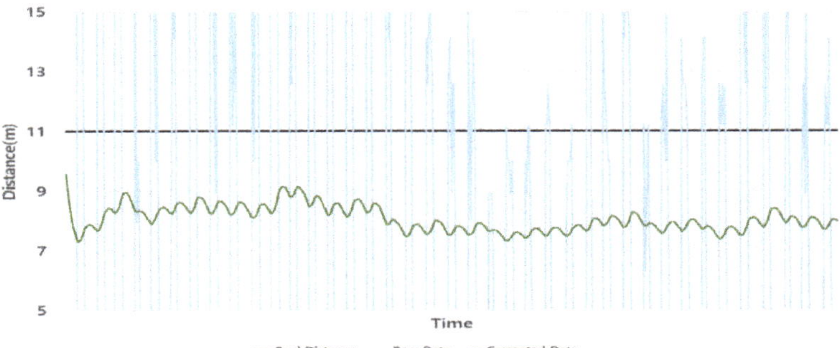

Figure 11. Distance measurement result (11 m).

5.3. Indoor Localization

This section describes the calculation of the position coordinates and analyzes the performance of the proposed indoor positioning system. Figure 12 depicts the experimental setup of the measurement space environment. There are many beacons spaced 8 m apart in the indoor space where the measurement takes place. The process began when a user's mobile device received a beacon broadcasting signal. If three or more signals were received at the same time, three nearby values were selected, and the distances from the corresponding beacons were measured. After applying the filter algorithm and calculating the distance from each beacon, the position coordinates of the mobile device were determined using the trilateration method.

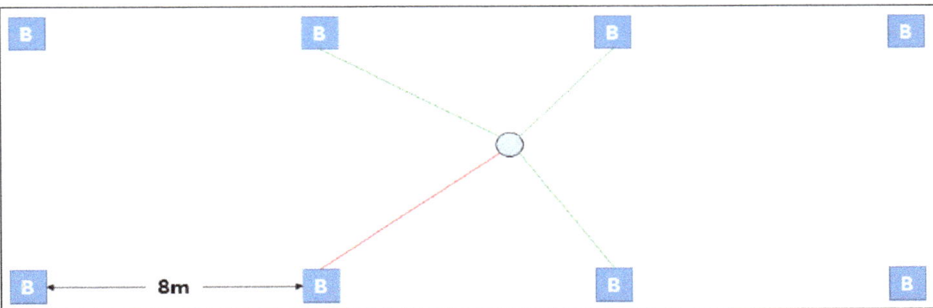

Figure 12. Indoor positioning environment.

Figure 13 shows the result of coordinate the smart device in the experimental space. The experimental space is 8 m × 8 m and beacons are installed at each corner. In order to analyze the performance, several measurements were made from specific locations to several devices and the result were analyzed. As observed Figure 13, the result of using raw data is unstable. Thus, coordinates are recorded at different position every time and are different from the actual position.

In contrast, the data that applied the filter algorithm was self-reinforcing and confirmed that the coordinates were measured at a certain position. We confirmed that the measured position differs by about 1m from the actual position.

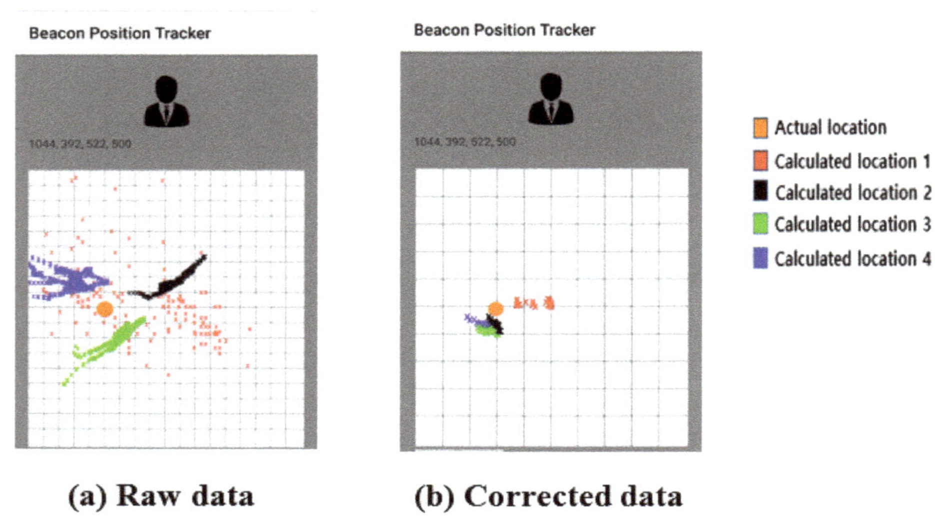

Figure 13. positioning results: (**a**) Raw data; (**b**) Corrected data.

6. Conclusions

In this paper, we propose an indoor positioning system that corrects unstable RSSI and improves the accuracy of location measurement by using filter algorithm based on average filter and EKF. In identifying beacon's identifier, we used ephemeral ID to improve security. We used Google's Eddystone-EID frame to generate ephemeral ID. We calculate the distance between the beacon and the device using the filter algorithm applied data at various distance and compared the results with the raw data. And we used the data to coordinate the position of the device and analyze the results. As a result, we confirmed that our method can achieve stable performance at distances of less than 7 m. However, performance degrades in a space where the distance between the beacon and the device exceeds 7 m. In the future, we will study algorithms that can achieve good performance even when the distance between the beacon and the device is farther. We intend to develop an indoor positioning system with good performance in a wider space even with a small number of beacons.

Author Contributions: J.K. and J.S. conceived the algorithm and designed the system. J.K. performed the experiments; J.S. analyzed the data; Y.W. guided this work and provided extensive revisions during the study. All authors have read and approved the final manuscript.

Funding: This research received no external funding.

Acknowledgments: This research was supported by the MSIT (Ministry of Science and ICT), Korea, under the ITRC (Information Technology Research Center) support program (IITP-2018-2016-0-00304) supervised by the IITP (Institute for Information & communications Technology Promotion).

Conflicts of Interest: The authors declare no conflict of interest.

References

1. KISA. LBS Industry Trend Report. Available online: http://www.kisa.or.kr/uploadfile/201805/201805281657080943.pdf (accessed on 15 June 2018).
2. Technavio. Global Location-Based Services (LBS) Market 2017–2021. Available online: https://www.prnewswire.com/news-releases/global-location-based-services-lbs-market---analysis-technologies--forecasts-to-2021---growing-popularity-of-proximity-based-mobile-advertising-among-enterprises---research-and-markets-300465016.html (accessed on 17 June 2018).

3. Mercer, D. Connected World: The Internet of Things and Connected Devices in 2020. Available online: https://www.strategyanalytics.com/access-services/devices/connected-home/consumer-electronics/reports/report-detail/connected-world-the-internet-of-things-and-connected-devices-in-2020 (accessed on 20 June 2018).
4. Inoue, Y.; Sashima, A.; Kurumatani, K. *Indoor Positioning System Using Beacon Devices for Practical Pedestrian Navigation on Mobile Phone*; Springer: Berlin/Heidelberg, Germany, 2009; pp. 251–265.
5. Lin, X.-Y.; Ho, T.-W.; Fang, C.-C.; Yen, Z.-S.; Yang, B.-J.; Lai, F. A mobile Indoor Positioning System Based on iBeacon Technology. In Proceedings of the 37th Annual International Conference of the IEEE Engineering in Medicine and Biology Society (EMBC '15), Milan, Italy, 25–29 August 2015; pp. 4970–4973.
6. Rida, M.E.; Liu, F.; Jadi, Y.; Algawhari, A.A.A.; Askourih, A. Indoor location position based on Bluetooth Signal Strength. In Proceedings of the 2nd International Conference on Information Science and Control Engineering, Shanghai, China, 24–26 April 2015.
7. Momose, R.; Nitta, T.; Yanagisawa, M.; Togawa, N. An accurate indoor positioning algorithm using particle filter based on the proximity of Bluetooth beacons. In Proceedings of the IEEE 6th Global Conference on Consumer Electronics (GCCE), Nagoya, Japan, 24–27 October 2017.
8. Pelant, J.; Tlamsa, Z.; Benes, V.; Polak, L.; Kaller, O.; Bolecek, L.; Kufa, J.; Sebesta, J.; Kratochvil, T. BLE device indoor localization based on RSS fingerprinting mapped by propagation modes. In Proceedings of the 2017 27th International Conference Radioelektronika (RADIOELEKTRONIKA), Brno, Czech Republic, 19–20 April 2017; pp. 1–5.
9. Luo, W.; Deng, X.; Zhang, F.; Wen, Y.; Kadhim, D.J. Positioning and guiding educational robots by using fingerprinting of WiFi and RFID array. *EURASIP J. Wirel. Commun. Netw.* **2018**, *2018*, 170. [CrossRef]
10. Dong, X.; Se, B.; Jiang, R. Indoor robot localization combining feature clustering with wireless sensor network. *EURASIP J. Wirel. Commun. Netw.* **2018**, *2018*, 175. [CrossRef]
11. Github. Google/Eddystone: Eddystone-UID. Available online: https://github.com/google/eddystone/tree/master/eddystone-uid (accessed on 13 July 2018).
12. Github. Google/Eddystone: Eddystone-URL. Available online: https://github.com/google/eddystone/tree/master/eddystone-url (accessed on 13 July 2018).
13. Github. Google/Eddystone: Eddystone-TLM. Available online: https://github.com/google/eddystone/tree/master/eddystone-tlm (accessed on 13 July 2018).
14. Github. Google/Eddystone: Eddystone-EID. Available online: https://github.com/google/eddystone/tree/master/eddystone-eid (accessed on 13 July 2018).
15. Github. Google/Eddystone: Eddystone-EID Computation. Available online: https://github.com/google/eddystone/blob/master/eddystone-eid/eid-computation (accessed on 13 July 2018).
16. Shchekotov, M. Indoor Localization Method Based on Wi-Fi Trilateration Technique. In Proceedings of the 16th Conference of Fruct Association, Oulu, Finland, 27–31 October 2014.
17. Kwak, J.; Sung, Y. Beacon-Based Indoor Location Measurement Method to Enhanced Common Chord-Based Trilateration. *J. Inf. Process. Syst.* **2017**, *13*, 1640–1651. [CrossRef]
18. Github. Google/Eddystone/Configuration-Service/Eddystone Configuration GATT Service. Available online: https://github.com/google/eddystone/tree/master/configuration-service (accessed on 13 July 2018).
19. Welch, G.; Bishop, G. *An Introduction to the Kalman Filter*; University of North Carolina at Chapel Hill: Chapel Hill, NC, USA, 2001.
20. Paul, A.S. RSSI-Based Indoor Localization and Tracking Using Sigma-Point Kalman Smoothers. *IEEE J. Sel. Top. Signal Process.* **2009**, *3*, 860–873. [CrossRef]

Article

A Q-Learning-Based Approach for Deploying Dynamic Service Function Chains

Jian Sun [1], Guanhua Huang [1], Gang Sun [1,2,*], Hongfang Yu [1,2], Arun Kumar Sangaiah [3] and Victor Chang [4]

1 Key Lab of Optical Fiber Sensing and Communications (Ministry of Education), University of Electronic Science and Technology of China, Chengdu 611731, China; sj@uestc.edu.cn (J.S.); ghh211911@gmail.com (G.H.); yuhf@uestc.edu.cn (H.Y.)
2 Center for Cyber Security, University of Electronic Science and Technology of China, Chengdu 611731, China
3 School of Computing Science and Engineering, Vellore Institute of Technology, Tamil Nadu 632014, India; arunkumarsangaiah@gmail.com
4 International Business School Suzhou (IBSS), Xi'an Jiaotong-Liverpool University, Suzhou 215123, China; victorchang.research@gmail.com
* Correspondence: gangsun@uestc.edu.cn; Tel.: +86-28-61831578

Received: 12 October 2018; Accepted: 14 November 2018; Published: 16 November 2018

Abstract: As the size and service requirements of today's networks gradually increase, large numbers of proprietary devices are deployed, which leads to network complexity, information security crises and makes network service and service provider management increasingly difficult. Network function virtualization (NFV) technology is one solution to this problem. NFV separates network functions from hardware and deploys them as software on a common server. NFV can be used to improve service flexibility and isolate the services provided for each user, thus guaranteeing the security of user data. Therefore, the use of NFV technology includes many problems worth studying. For example, when there is a free choice of network path, one problem is how to choose a service function chain (SFC) that both meets the requirements and offers the service provider maximum profit. Most existing solutions are heuristic algorithms with high time efficiency, or integer linear programming (ILP) algorithms with high accuracy. It's necessary to design an algorithm that symmetrically considers both time efficiency and accuracy. In this paper, we propose the Q-learning Framework Hybrid Module algorithm (QLFHM), which includes reinforcement learning to solve this SFC deployment problem in dynamic networks. The reinforcement learning module in QLFHM is responsible for the output of alternative paths, while the load balancing module in QLFHM is responsible for picking the optimal solution from them. The results of a comparison simulation experiment on a dynamic network topology show that the proposed algorithm can output the approximate optimal solution in a relatively short time while also considering the network load balance. Thus, it achieves the goal of maximizing the benefit to the service provider.

Keywords: network function virtualization; service function chain; reinforcement learning; load balancing; security

---

1. Introduction

Currently, most networks use a large number of dedicated hardware devices that provide features such as firewalls and network address translation (NAT). The various services provided by service providers usually require specialized hardware devices. As the network grows in size and emerging industries such as big data [1,2] and cloud computing [3–5] are expanding rapidly, starting a new service requires deploying a variety of dedicated hardware devices, making it extremely difficult to design a private protocol and protect the security of user data. Because users share resources on dedicated hardware devices, hackers can take advantage of certain security vulnerabilities in some devices and easily obtain data of users. Nevertheless, as service requirements continue to increase, service providers must regularly expand their physical infrastructure, leading to high infrastructure and operating costs [6,7].

Based on network service requirements, service providers have begun to pay attention to network function virtualization (NFV) technology [8]. Unlike previous dedicated hardware devices, NFV converts network service functions into software, and provides services through a common server device. Although using the same servers, NFV technology isolates the resources and functions on the computer and then allocates them to users, which ensures the privacy and security of user data. Using NFV, service providers can respond quickly to service needs and traffic changes by managing software to distribute services. Moreover, by centralizing resource management, service providers can reduce infrastructure and operating costs [9].

Using NFV technology, we can use virtual network functions (VNFs) [10] to represent network services deployed on continuous network topology nodes to form a service function chain (SFC) [11]. This approach reduces hardware costs and operating expenses but needs to be robust for IT and bandwidth resources [12]. However, the development of NFV also introduces challenges. For example, as user demand grows, the SFC that provides services to the user may need to be adjusted frequently to ensure service continuity, for example, when the user moves. In this case, preserving continuity requires changing the SFC in the network topology to suit the user's requests [13]. However, an intelligent approach to building and adjusting SFCs to reduce labor costs is worth investigating.

For an SFC, when a service ends, the best deployment for a new request may change [14]. Static SFC deployment causes unnecessary resource consumption and wastes many idle resources. Therefore, dynamic SFC deployment is more suitable for research; we must balance the consumption of various resources in real time.

Reinforcement learning (RL) [15] is an area of machine learning that focuses on determining how software agents should act in an environment to maximize some of the concepts of cumulative rewards. This problem is broad and applicable to many fields [16–20].

RL is also used to study the existence of optimal solutions and accurate algorithm calculation to adapt to and explore unknown environmental models without requiring training data support. In some cases, RL can be used to yield limited-equilibrium decisions. RL refers to an agent in an environment that involves many states. Through RL, each agent learns appropriate actions by receiving rewards for actions taken to achieve its purpose. The states that exist in an environment may or may not help the agent achieve the goal. As agents experience each state while attempting to achieve their goals, they are rewarded by the environment. When an agent consistently fails to reach the target state, it cannot gain the reward. Thus, agents iterate through many trials and errors and eventually learn the most appropriate action for each state.

However, there are many researches related to reinforcement learning, such as stochastic learning automata [21,22], Q-learning [23], deep Q network [24], and Generative Adversarial Networks [25]. After our comparison, we found that Q-learning is most suitable for the problems studied in this paper.

In this study, we use an RL algorithm called Q-learning. The rewards and strategies are recorded in a matrix. The goal of the training phase is to make the strategy that preserves the matrix (*Q* matrix) converge. The matrix has been proven to converge for continuous decision problems in environments that meet the requirements of RL. During use, the state transition strategy is provided directly by the *Q* matrix containing the policy.

The reasons why we use Q-learning are as follows. First of all, because a common problem of machine learning is it is time consuming in training; the Q-learning training stage is easy to understand, and the structure of the storage strategy is easy to be modified, so it can be optimized. Secondly, the problem that Q-learning can solve is more similar to the problem in this paper, which helps to give play to the advantages of the algorithm. Therefore, the use of the Q-learning algorithm can greatly reduce the training time and computational complexity.

To optimize the deployment of SFCs in a dynamic network, we integrated RL into the problem and designed a new deployment decision algorithm. This paper studies the problem of deploying SFCs in a multiserver dynamic network. Unlike the data center network [26,27], the nodes of the multi-server network have fewer resources and the SFC deployment is more difficult. Due to the characteristics of dynamic networks, new SFCs may need to be deployed at any time, and some services should be cancelled. To accomplish these tasks, we propose a real-time online deployment decision algorithm called QLFHM. After learning the entire topology and the use of virtual resources, the algorithm uses the RL module and the load balancing module to output an SFC immediately. We compared our proposed algorithm with other algorithms in a simulation experiment and evaluated it repeatedly. The simulation results show that the algorithm achieves good performance with regard to decision time, load balancing, deployment success rate and deployment profit.

The rest of this paper is organized as follows. Section 2 provides an overview of the current work related to the field. In Section 3, we describe the problem models we want to solve, including network models, user requests, and dynamic deployment adjustments. To solve these problems, we propose our algorithm model in Section 4. We present a comparison with other algorithms in Section 5. Finally, Section 6 summarizes the paper.

2. Related Work

In NFV networks, network functions are implemented as VNFs in software form. The characteristics of VNFs allow them to be deployed flexibly and ensure the security of users. Therefore, key consideration needs to be given to the placement of VNFs to meet service requirements, quality of service, and the interests of service providers. This type of problem is called the VNF Placement (VNF-P) problem and has been proven to be a non-deterministic polynomial-time hard (NP-hard) problem [28]. Consequently, it is often difficult to find the optimal solution of a VNF-P problem.

The study of deployment problems is divided into static deployment problems and dynamic deployment problems. The difference is that during static deployment, the SFC in the network is always there; in contrast, during dynamic deployment it will be withdrawn after some period.

In a static problem, deployment is the equivalent of an offline decision: all the requirements are considered when choosing the deployment. Because the SFC being deployed is not retracted after placement, the main consideration is how to arrange more SFCs, which is also the main evaluation criterion. For example, the BSVR algorithm proposed by Li et al. [29] mainly considers load balancing and the number of accepted SFCs. In addition, unlike us, they set up a consistent type of VNF that can be shared by multiple SFCs.

Here, we study the dynamic problem, which is closer to the real network situation [30]. In a dynamic situation, an SFC will be withdrawn after some deployment period, making the network more fluid.

Facing those problems, the methods are similar. To obtain the optimal solution of the VNF-P problem, mathematical programming methods such as integer linear programming (ILP) and mixed ILP (MILP) are the most popular approach [31]. The next most popular approaches involve heuristic algorithms [32] or a combination of heuristic algorithms and ILP. Although there are different optimization approaches to VNF-P problems, the limitations of these approaches are generally similar and include bandwidth resources, IT resources, link delay, VNF deployment, and cost and profit considerations [33–35].

For example, Bari et al. [28] approximately expressed the VNF-P problem as an ILP model and solved it with a heuristic algorithm that attempted to minimize operating expense (OPEX) and maximize network utilization. Gupta et al. [33] tried to minimize bandwidth consumption. Luizelli et al. [31] also developed an ILP model that seeks to minimize both the end-to-end delay and the resource overhang ratio. J. Liu et al. [14] proposed the column generation (CG) model algorithm based on ILP and attempted to maximize the service provider's profit and the request acceptance ratio.

Some of the papers mentioned above have reported that the execution times for solving these two mathematical models increase exponentially with the size of the network. After solving the model with optimization software or a precision algorithm, they immediately proposed a corresponding heuristic algorithm.

Although the execution times of heuristic algorithms is much lower than that of ILP, most existing heuristic algorithms provide only near-optimal solutions. However, considering the time savings, heuristic algorithms form the main approach to solving VNF-P problems.

Some recent solutions to the VNF-P problem have applied machine learning techniques. Kim et al. [36] constructed the entire problem as an RL model. Although the results of this approach may not differ much from the optimal solution, using it in complex network situations results in extremely long training times.

We also tried to avoid the shortcoming of too long training time while using intensive learning. Given the tradeoff between the accuracy of the ILP algorithm and the time efficiency of the heuristic algorithm, this paper proposes a QLFHM algorithm that combines RL and heuristic algorithms. After comparing QLFHM with benchmark algorithms, we conclude that the QLFHM algorithm not only guarantees an approximately optimal solution but also guarantees the time efficiency when dynamically deploying a SFC.

3. Problem Description

We studied the problem of deploying an SFC across multiple servers in a dynamic network. Our goal is to make the service provider most profitable while providing security guaranteed services.

We consider a scenario in which multiple input requests need to be deployed from the source server to the target server over an appropriate link. The link must support the VNFs included in the request. Due to the limited capacity of all servers and links, consideration should be given to the distribution of SFCs to be deployed for the request that will allow more SFCs to be deployed.

We describe the problem model in the next sections, including the research motivation, network model, request model and dynamic SFC deployment.

3.1. Research Motivation

Given a network with multiple servers, and each server supports the deployment of VNFs, but IT resources are limited. Link bandwidth resources are also limited. Requests dynamically switch between active and offline states. Thus, effectively determining how to deploy the SFCs can maximize the request acceptance ratio and the service provider's profit and minimize the computation time, while satisfying all the constraints.

3.2. Network Model

The network can be seen as a graph $G = (V, E)$, where V denotes the set of nodes, and E is the set of links between nodes. Each $e \in E$ represents a physical link between two network nodes; we use $B_e \in N$ to show a link's bandwidth capacity. Each $v \in V$ is a network server, which functions as both the users' access point and a switch; each server also has an IT resource capacity; we use $I_v \in N$ to denote the IT resources of v. T represents the set of all the VNFs. We use $T_v \in T$ to denote the set of VNFs that can be deployed at each v. All servers can offer NFV services; however, some servers only support partial services. We assume that bandwidth resources and node computing resources are limited. We use a Boolean variable $p_{j,v}$ to represent the state in which the j-th VNF vnf_j is deployed on $v \in V$. A $p_{j,v}$ value of 1 denotes deployable and a value of 0 denotes undeployable:

$$p_{j,v} = \begin{cases} 1\, if\ vnf_j \in T_v\,, I_v > 0 \\ 0\, else \end{cases} \tag{1}$$

3.3. Request Model

RE is used to represent all incoming requests. Each request $i \in RE$ is represented by the following variables: s_i, d_i, P_i, r_i. Here, $s_i \in V$ refers to the client's access node, $d_i \in V$ is the data provider node required by the user, $P_i \in T$ is a vector that includes the required VNFs sequence on the request SFC, and $r_i \in R^+$ refers to the unit compensation paid after the successful deployment of the request, which is related to the number of VNFs represented by num_vnfs_i. We use ω to represent the unit value. For convenience, we make $\omega = 1$:

$$r_i = \frac{3}{2} * num\_vnfs_i * \omega. \tag{2}$$

The profit gained after successful deployment of the SFC of user i is represented by $profit_i$, and l_i represents the chain length of a successfully deployed SFC: the number of nodes in the SFC:

$$profit_i = r_i - l_i * \omega. \tag{3}$$

A successfully deployed sfc_i should match the starting point s_i and destination d_i, select an appropriate chain, and arrange the VNFs sequence sequentially on the chain nodes. The link length l_i is limited by the compensation r_i. Service providers need to ensure their profitability:

$$l_i < \frac{r_i}{\omega}. \tag{4}$$

x_i is a Boolean variable that indicates whether the request of user i is successfully deployed. If its SFC is online, x_i is 1; otherwise, x_i is 0. $Node\_SFC_i$ represents all the nodes in sfc_i. $Node\_vNFs_i$ represents the nodes that can deploy VNFs in sfc_i. $Link\_SFC_i$ represents all the links in the sfc_i. $z_{i,j,\pi}$ is a Boolean variable that equals 1 if user i uses the path $\pi \in Link\_SFC_i$ and the next node that deploys VNFs of its node-deployed j-th VNF is node deployed—the $(j + 1)$-th VNF—and 0 otherwise.

Equation (5) ensures that the nodes deploying VNFs do not include the user access node s_i or the service access node d_i:

$$Node\_vNFs_i = Node\_SFC_i - (s_i + d_i). \tag{5}$$

Equations (6) and (7) ensure that when the sfc_i is online, the P_i is deployed in the sfc_i in sequence:

$$\bigcap_{v \in Node\_vNFs_i} p_{j,v} = x_i, \forall vnf_j \in P_i \, \forall i \in RE \tag{6}$$

$$\bigcap_{\pi \in Link\_SFC_i} z_{i,j,\pi} = x_i \, , \forall vnf_j \in P_i \, \forall i \in RE. \tag{7}$$

3.4. Dynamic SFC Deployment

We assume that during the arrival of a dynamic request scenario, the service provider will provide service for the new request and will cancel the service function of a previous SFC at the end of the service request time. The arrival time of requests occurs at a certain time interval. Therefore, at every moment, the service provider addresses two types of user requests. These mainly affect the service provider's operation expenses and involve checking whether a new request is available and whether there a chain of online services exists that need to be cancelled.

The goal of dynamic deployment is to maximize service provider profits. The IT resources and bandwidth capacity exist as the deployment constraint condition; however, they affect only the deployment ability and not the operation cost.

$Node\_vNFs\_put_i$ represents the node set that deployed VNFs for sfc_i. I_max_v represents the maximum IT capacity of node v. I_use_j means the IT capacity needed by j-th VNF. $Link\_vNFs\_put_i$ is the set of links that belong to sfc_i. B_max_e denotes the maximum bandwidth capacity of link e, and B_use_i is the bandwidth capacity needed by the sfc_i.

For every $v \in V$:

$$\sum_{i \in RE} I_{usej} * p_{j,v} * x_i \leq I\_max_v \text{ , } if\ v \in Node\_vNFs\_put_i, \forall vnf_j \in P_i, \tag{8}$$

and for every $e \in E$:

$$\sum_{i \in RE} B\_use_i * x_i \leq B\_max_e \text{ , } if\ e \in Link_{vNFs_{put_i}} \tag{9}$$

Equations (8) and (9) ensure that the user will not use more bandwidth and IT resources than the total capacity available during any service time.

It should be noted, however, that some requests may be blocked because their long deployment chains result in no profits. The Boolean variable y_i indicates whether the request of user i has been successfully deployed. The goal of this problem is to maximize the service provider's profit K, described as follows:

$$K = \sum_{i \in RE} profit_i * y_i. \tag{10}$$

We depict the dynamic SFC deployment and revocation process in Figure 1. At each moment, the situations represented by Figure 1a,b may occur in the network.

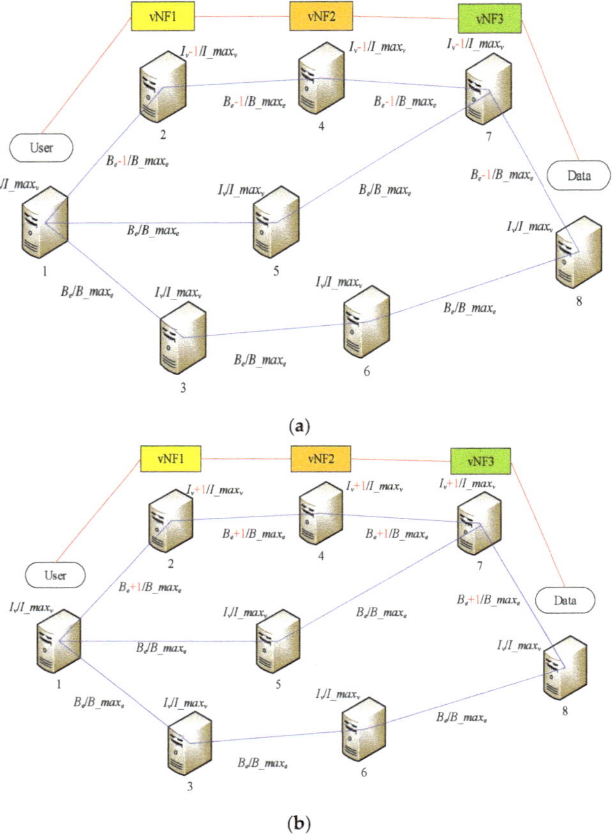

Figure 1. (**a**) Service function chain (SFC) deployment; (**b**) SFC revocation; SFC deployment and revocation.

In Figure 1a, after the SFC is generated for a request, the SFC will be deployed to the corresponding path when sufficient resources are available. The path's bandwidth resources will be consumed, assuming a unit is consumed. The IT resources of the server deploying VNF will also be consumed, assuming a unit is consumed, but the user access node and the data provider node do not consume IT resources.

In Figure 1b, after the service time of an SFC expires, the SFC will be dropped from the network. The path's bandwidth resources and the IT resources of the server that deployed the VNF will also be recovered, if they used units.

To maximize profits, service providers should first attempt make the SFC shorter while meeting the demand. For the overall network, IT resources and bandwidth resources must be balanced properly, which means that SFCs should be distributed as widely as possible, rather than crowding them together, which can create problems. In this way, at any given time, we will have more nodes and paths to choose from to form new SFCs.

4. Q-Learning Framework Hybrid Module Algorithm

In this section, we use the Q-learning framework hybrid module algorithm (QLFHM) to address dynamic SFC deployment. First, to reduce the problem complexity, we divide the solution process into two parts: we use the RL module to output several of the shortest paths that meet certain requirements. The load balancing module obtains multiple routing outputs from the previous module and finally obtains the solution of the problem. The goal of hierarchical processing is to reduce the training and learning times and achieve an efficient output scheme. The architecture of QLFHM is shown in Figure 2.

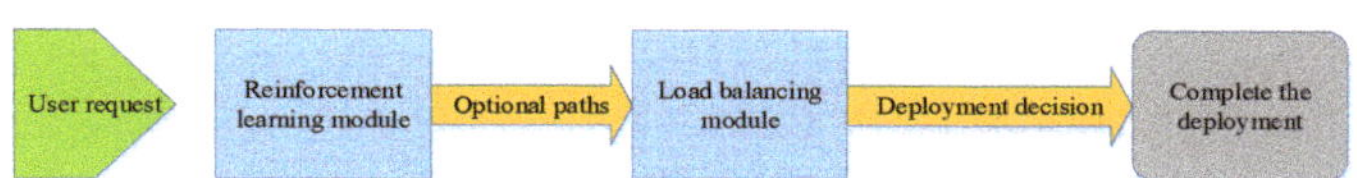

Figure 2. Q-learning Framework Hybrid Module.

4.1. Preliminaries

Problems that can be solved by Q learning generally conform to the Markov decision-making process (MDP) and have no aftereffect. That is to say, the next state of the system is only related to the current state information and is unrelated to the earlier state. Unlike the Markov chain and Markov models, the MDP considers actions, that is, the next state of the system is related not only to the current state, but also to the current action taken. The dynamic process of MDP is shown in Figure 3:

Figure 3. Dynamic process of the Markov decision-making process (MDP).

The return value *r* is based on state *s* and action *a*, each combination of *s* and *a* has its own value of return, then MDP can also be represented as the following Figure 4:

In relation to the problem in this paper, it conforms to the Markov decision making process, but the difference is that the times of decision making for the problem in this paper is limited. Therefore, after our study, we combine the Q-learning with this problem and optimize the Q-learning algorithm for this problem. In this way, we can not only take advantage of the reinforcement learning, but also avoid its defects, which can provide new ideas for dynamic SFC deployment.

A key point of the algorithm proposed in this paper is that it improves the *Q* matrix and changes the original two-dimensional matrix into a five-dimensional matrix. The five subscripts are *now_h*, *now_node*, *action_node*, *end_node*, *h*. The subscript *now_h* refers to the number of hops that have been visited in the current state; *now_node* is the node in which the agent is in the current state; *action_node* is the next available node set; *end_node* is the node that this SFC will eventually reach; and h is the minimum number of hops that can meet the deployment requirements.

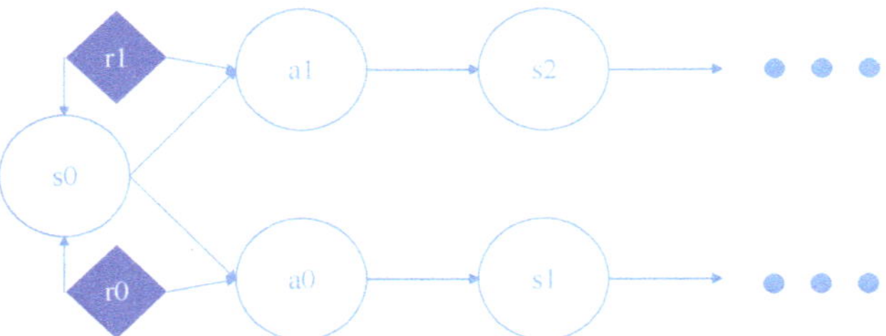

Figure 4. Diagram of r with a and s.

As shown in Figure 5, after the agent responds to the environmental state, the environmental state changes and returns r. The Q matrix is updated with r. The agent will repeat the above behavior until the Q matrix converges. The Q-learning algorithm is shown in Equation (11):

$$Q(s,a) = Q(s,a) + \alpha\left(r + \gamma \max_{a'} Q(s',a') - Q(s,a)\right). \tag{11}$$

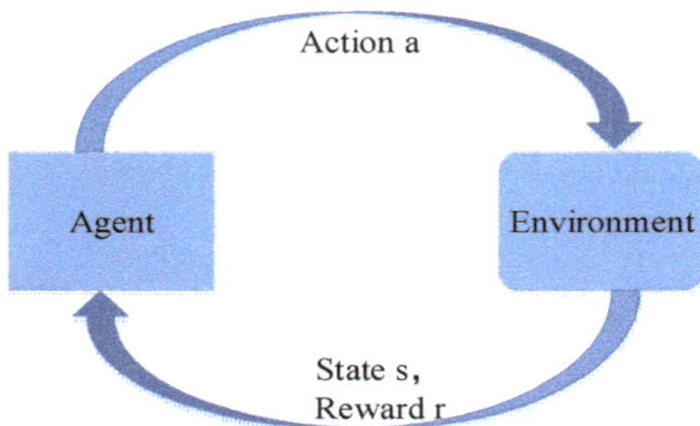

Figure 5. Reinforcement learning training stage.

The values stored in Q are the recommended values for the next action in each state. The higher a recommended value is, the more it is worth performing. The recommended values are formed during the training phase of the Q matrix.

In Equation (11), Q is a matrix that stores the recommended values of the executable actions in the current state. Depending on these values, the agent can decide which action to take next. In the Q matrix, the subscript s refers to the state, a to the action, $s\prime$ to the future state, $a\prime$ to the future action, and r is the reward value, which comes from the reward matrix R. Here, α and γ are the studied ratio between 0 and 1.

In R, we set the value of the element to 1000, which has same *now_node* and *end_node* in the subscript. This value represents the reward given when completing the pathfinding task.

For Q [*now_h*, *now_node*, *action_node*, *end_node*, *h*], Specifically, when the four subscripts *now_h*, *now_node*, *end_node*, and *h*, which represent states, are determined, the subscripts of *action_node* are iterated. The maximum value is selected and its subscript *action_node* is executed. The advantages of this approach are that (1) it makes the state more observable, and (2) it divides the states into several independent parts that are conducive to parallel programming techniques.

The recommended values stored in Q are determined by Equation (11), which is a simplified version of Equation (12).

$$Q(s,a) = \alpha\left(r + \underset{a'}{max}Q(s',a')\right). \tag{12}$$

Some of the parameters and variables used in the QLFHM algorithm are described in Table 1:

Table 1. Parameters and variables in the Q-learning Framework Hybrid Module algorithm (QLFHM) algorithm.

Parameters and Variables	Definition
G	Information about network topology
V	A list of nodes in a network topology
V_i	Node adjacent to node i
h_{max}	The maximum number of hops allowed by the model
h_{min}	The minimum number of hops allowed by the model
Q	A 5-dimensional matrix with 5 subscripts $\{now\_h, now\_node, action\_node, end\_node, h\}$ that store recommended action values
R	A 5-dimensional matrix with 5 subscripts $\{now\_h, now\_node, action\_node, end\_node, h\}$ that store action reward values
h	The number of hops in the current state
RE	User request list, including start and stop nodes, VNF requirements, arrival time, and request online duration
PA	Path list that matches the start and stop nodes
CA	Path list that satisfies the start and stop nodes and the VNFs requirements
ONL	The online SFC list

4.2. *Reinforcement Learning Module*

In this section, we propose the RL module, which is responsible for outputting alternative paths based on the network topology. The content is divided into two parts: a training stage and a decision stage. In the first part, we first provide the original Q-learning training algorithm and then provide the optimized training algorithm for this problem. Both algorithms have advantages and disadvantages. In the second part, we will propose the algorithm used in the decision stage.

4.2.1. Original Q-Learning Training Algorithm

In the training phase, training data are not required; this phase automatically generates the RL model according to the basic network topology information.

Algorithm 1 adopts the standard Q-learning training method, that is, iterative trial and error. The rewards attained during the repeated attempts finally cause the Q matrix to converge. The variable u is added for the greedy algorithm that enables the agent to improve the explored path in most cases, while making it possible to explore new paths.

The advantages of using the Q-learning algorithm in this paper are as follows: (1) we can observe and understand the decision-making process, and the use of the Q matrix is more intuitive and comprehensible; (2) we can quickly convert this algorithm to a deep Q-learning algorithm, which uses a neural network (DQN) to replace the Q matrix for decision making; and (3) Q-learning is conducive to the improvement of the algorithm proposed in this paper and is suitable for solving the problems in this paper.

The algorithm based on Q-learning obtained satisfactory results, but it also faces some problems. For example, in the face of complex situations or too-large networks, the training period will be excessive. Consequently, an improved version is presented in Algorithm 2, which is optimized for our problem.

Algorithm 1. Original Q-learning Training Algorithm

```
initialize the Q matrix with all zero elements;
initialize the R matrix;
initialize h_min and h_max;
h = 0;
While True do
   randomly generate v1 ∈ V;//as the now_node
   randomly generate v3 ∈ V;//as the end_node
   For h < h_max do:
      randomly generate u ∈ [0, 1];
       If u ≤ 0.8 then:
          choose the v2 ∈ V_v1 with the maximum recommended value from Q;
       End If
       If u > 0.8 then:
          randomly choose a v2 ∈ V_v1;
       End If
       v1 = v2;
       If v2 = v3 then:
          Write the link to the Q matrix using Equation (12);
          Break;
       End If
       h++;
       If h = h_max then:
          Break;
       End If
    End For
    If the Q matrix has basically converged, then:
       Break;//return the Q matrix that can be used
    End If
 End While
```

4.2.2. Optimized Q-Learning Training Algorithm

Algorithm 2 is an improved version of the Q-learning algorithm based on our problem. It abandons the trial-and-error learning mode of the original algorithm and adopts a method similar to neural diffusion, which results in a hundredfold reduction in training time.

In the Q matrix, we did not list the state with one index as in the original algorithm of Q-learning; instead, we divided the state into four indexes. The advantages of this approach are as follows: Algorithm 2 normally executes on a single computer; however, when greater efficiency is required, the algorithm can begin working in a distributed operation starting on line 5. Because the four indexes make some states independent, we can use distributed computing to reduce the execution time. And the main functions in Algorithm 2 are described in Algorithm 3.

Algorithm 2. Optimized Q-learning Training Algorithm

```
1: initialize the Q matrix with all zero elements;
2: initialize the R matrix;
3: initialize h_min and h_max;
4: h = 0;
5: For each node v ∈ V      do //as the end_node
6:    chain = [v]
7:    Find_way (Q, R, G, h_min, h_max, h, chain)
8: End For
```

Algorithm 3. Find_way ($Q, R, G, h_{min}, h_{max}, h, chain$)

$v0 = chain\,[0]$;
$h = h\,{+}{+}$;
$chain\_tmp = chain$;
While $h \le h_{max}$ **do**
  For each node $v2 \in V_{v0}$ **do**
    If $v2$ is not in $chain\_tmp$ **then**
      $chain\_tmp = v2 + chain\_tmp$;
      Find_way ($Q, R, G, h_{min}, h_{max}, h, chain$);
      If $h \ge h_{min}$ **then**
        For i in $chain\_tmp$ **do**
Write the link to the Q matrix,
$\mathrm{Q}(\mathrm{s}_i, \mathrm{a}_i) = 0.8\left(r + \underset{a_i'}{max} Q(s_i', a_i')\right)$
        End For
      End If
    End If
  End For
End While

4.2.3. Complexity Analysis of Original and Optimized Q-Learning Training Algorithm

In this section, we give the time complexity of the original and optimized Q-learning training algorithm.

We use *ite* to represent the number of iterations of the original Q-learning training algorithm in the trial and error training process, which is a very large number and also the reason for the long training time.

The time complexity of original Q-learning training algorithm is

$$\mathrm{T} = O(ite * h_{max}). \tag{13}$$

And the time complexity of optimized Q-learning training algorithm is

$$\mathrm{T} = O\left(n^{h_{max}}\right). \tag{14}$$

where h_{max} is less than 13; and n represents the number of nodes in the topology.

However, *ite* is not a constant, which will increase significantly with the increase of n and h_{max}. And Equation (14) shows the worst case of a full-connected-network. Therefore, for the problem solved in this paper, the optimized algorithm will consume much less time.

4.2.4. Q-Learning Decision Algorithm

After the Q matrix has largely converged, the training phase terminates. During the decision phase, we use the Q matrix to output multiple alternative paths CA that meet the input requirements. These paths are then sent to the load balancing module, which makes a final selection. The whole process is described in Algorithm 4.

It is worth mentioning that even in the decision stage, the Q matrix is not necessarily permanently static; it can be adjusted based on the actual situation to support supplementary learning for new paths or nodes or the removal of expired paths and nodes.

Algorithm 4. Q-learning decision-making process

read the trained matrix Q
read the user request list RE
For every re in RE **do**
  Select some optional paths PA from Q;
  For every pa in PA **do**
    If the pa can deploy the required VNFs **then**
      add pa to the candidate list CA;
    End If
  End For
  If the candidate list CA is empty **then**
    deployment for this re failed;
    continue;
  End If
  Send the candidate list CA to the load balancing module;
End For

4.3. Load Balancing Module

The load balancing module adopts a scoring system. It scores each SFC output from the previous module, and the optimal choice will be deployed.

First, consider the link weight $weight_{link} \in [0,1]$ and the node $weight_{node} \in [0,1]$, which represent a proportion that focuses on the link or node. When no special requirement exists, we set the weights to 0.5, 0.5:

$$weight_{node} + weight_{link} = 1. \tag{15}$$

Next, we consider the weights of specific nodes $weight_{node\ v} \in N$ and specific links $weight_{link\ e} \in$ N is considered. To urge the SFC to go through a node or link, we increase its weight, which increases the probability that the SFC will traverse that node or link. When no special requirement exists, the weights remain unchanged.

Finally, the link bandwidth resources B_e and node computing resources I_v are combined with the weights mentioned above to obtain the final score. The higher the score is, the more the path represented is worth deploying. The score is calculated by Equation (16):

$$score = weight_{link} * [\sum_{e \in SFC} (weight_{link\ e} * B_e)] + weight_{node} * [\sum_{v \in SFC} (weight_{node\ v} * I_v)]. \tag{16}$$

Using Equation (16), we can construct Algorithm 5, which takes the output of the RL module as input and outputs the final decision results.

Algorithm 5. The load balancing scoring process

read the information from G
read the candidate list CA
For every pa in CA **do**
  calculate the score of pa using Equation (15);
End For
take the path with the highest score from the candidate list CA;
record the start time $t_{start-re}$, and record the end time t_{end-re}
add re to the online SFC list ONL;
change the resource residuals in the topology;
If any re in ONL reaches t_{end-re} **then**
  return the related resources in the topology;
End If

Due to the flexibility of the independent scoring system, it can be customized for problems that involve required traversal nodes as well as nodes that need to be bypassed. By further adjusting parameters and structures, this algorithm can also be used to solve the problems related to virtual machine consolidation and dynamic application sizing [37].

Dividing the SFC dynamic network deployment problem into two parts reduces the scale of the problem and improves the execution efficiency. First, the improved Q-learning training algorithm results in a training time that is one hundred times smaller. In addition, the independent scoring system is highly flexible and can be customized for specific problems.

5. Performance Evaluation and Discussion

In this section, we compare the QLFHM algorithm with two other algorithms to evaluate the performance of the proposed dynamic SFC deployment method. We first describe the simulation environment and then provide several performance metrics used for comparisons in the simulation. Finally, we describe the main simulation results.

5.1. Simulation Environment

The simulation uses the US network topology, which has 24 nodes and 43 edges. Here, we assume that the server and switch are combined, which means that all nodes have local servers but not necessarily all VNFs. The server's IT resource capacity is 4 units, and the bandwidth capacity of each physical link is 3 units. Note that each VNF occupies 1 unit of IT resources, and each traversed link occupies only 1 unit of bandwidth resources. The online time of each request follows a uniform distribution, and the arrival time is subject to a Poisson distribution.

Some servers can support 5 VNF types, but not all servers support all VNF types. For each VNF type, the IT resources of each VNF consume one unit, and each unit can serve one user. Assume that the number of VNFs per user requested in SFC is normally distributed from [2–4]. To compare the proposed algorithm with existing algorithms, we implement the algorithm in [14], which has a high success rate due to its use of ILP. Although the algorithm is optimized for time, it still requires considerable time; thus, it is not shown on the time comparison graph. We also implement the algorithm in [28], which has good execution efficiency.

5.2. Performance Metrics

We used the following metrics in the simulation to evaluate the performance of our proposed algorithm. For the dynamic network with limited resources, we selected three sets of data for analysis: the request acceptance ratio, the average service provider profit, and the calculation time per request.

(1) **Request acceptance ratio:** This value is the ratio of incoming service requests that have been successfully deployed on the network to all incoming request. Ratio A is defined as

$$A = \frac{Number_{successfully\ deployed\ SFC}}{Number_{input\ service\ requests}}. \tag{17}$$

(2) **Average service provider profit:** This value is the total profit earned by the service provider after processing the input service requests. The average service provider profit K can be calculated as follows:

$$K = \sum_{i \in RE} \left(\frac{3}{2} * num\_vnfs_i - l_i \right) * \omega * y_i. \tag{18}$$

(3) **Calculation time per request:** This value reflects the decision time required before each SFC is deployed. The calculation time per request C is expressed as follows:

$$C = \frac{Total\ running\ time}{Number_{input\ service\ requests}}. \tag{19}$$

5.3. Simulation Results and Analysis

We divide the experiment into two parts and present and analyze the results separately.

The first part involves comparing and analyzing the performances of the QLFHM algorithm and the other two selected benchmark algorithms in the simulated network. The second part compares and discusses some parameters that can affect the performance of the QLFHM algorithm and demonstrate its flexibility and modular capabilities.

We obtained each data point by averaging the results of multiple simulations. We executed the simulations on an Ubuntu virtual machine running on a computer with a 3.7 GHz Intel Core i3-4170 and 4 GB of RAM. The algorithm models were coded in Python.

5.3.1. Performance Comparison in a Dynamic Network

This section provides comparison results from simulating the algorithm proposed in this paper and the two other selected algorithms [14,28] on an SFC dynamic deployment problem. Three sets of data were selected for analysis: the request acceptance ratio, service provider average profit, and the calculation time required for each request.

Figure 6 shows a comparison of the request acceptance ratio achieved by the three algorithms. As Figure 4 shows, when the number of requests is less than 400, the request acceptance ratio is unstable due to insufficient data. However, the QLFHM and CG algorithms always achieve better results than does the Viterbi algorithm. The request acceptance ratio of the CG algorithm is slightly different because more than one optimal path exists in some cases, and the two algorithms use different path selection strategies. After the algorithms select different paths, the overall situation will also differ, resulting in some overall differences. After the number of requests exceeds 400, the request acceptance ratio of the three algorithms tends to become stable; at that point, the request acceptance ratio of the QLFHM algorithm is roughly the same as that of the CG algorithm using ILP. This result demonstrates that the deployment success ratio of the QLFHM algorithm is higher than that of the Viterbi algorithm and is close to the optimal solution at any request scale.

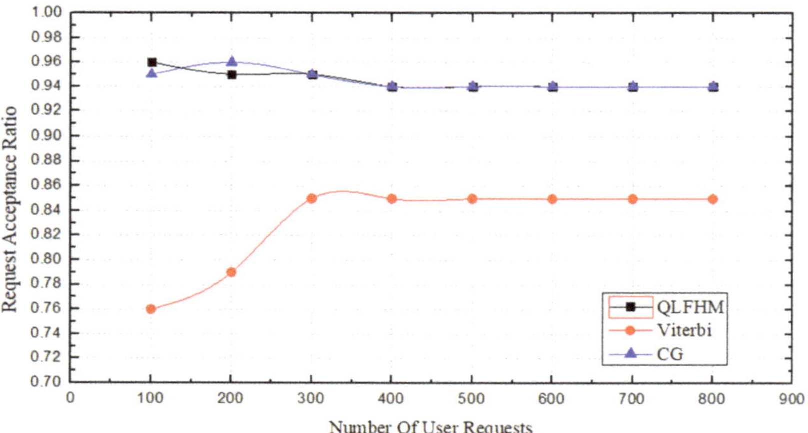

Figure 6. Comparison of the request acceptance ratio.

Figure 7 shows a comparison of the profits of the service providers achieved by the three algorithms. There is almost no profit difference between the QLFHM and CG algorithms. However, as the number of requests increases, the profit difference between the QLFHM algorithm and the Viterbi algorithm gradually increases. Because the CG algorithm uses ILP, its deployment scheme is close to optimal. The QLFHM algorithm obtains results not much different from CG, indicating that the deployment scheme of the QLFHM algorithm is close to optimal. We are confident that under larger numbers of requests, the profit obtained using the QLFHM algorithm will never be less than that obtained by the other two algorithms.

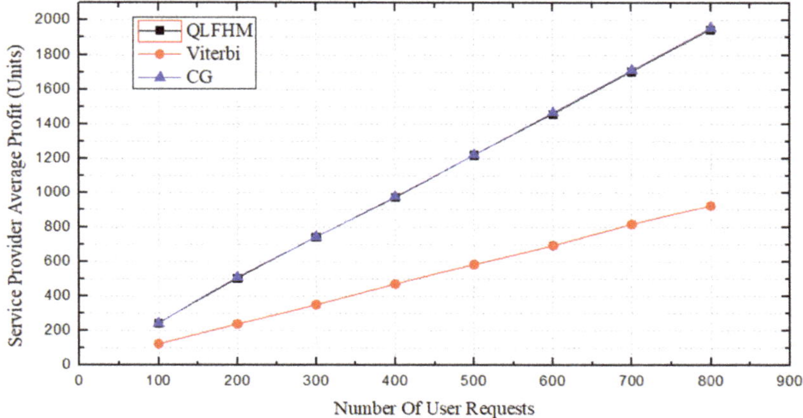

Figure 7. Comparison of service provider average profit.

In Figure 8, we compare the average operation time of only two algorithms. We know from [14] that the operation time of the CG algorithm is much longer than that of the other two; therefore, the CG algorithm's performances are not included in the figure. Figure 8 shows that the average operation time of the QLFHM algorithm is approximately 6 times less than that of the Viterbi algorithm. This result indicates that among the three algorithms, the QLFHM algorithm yields the fastest result.

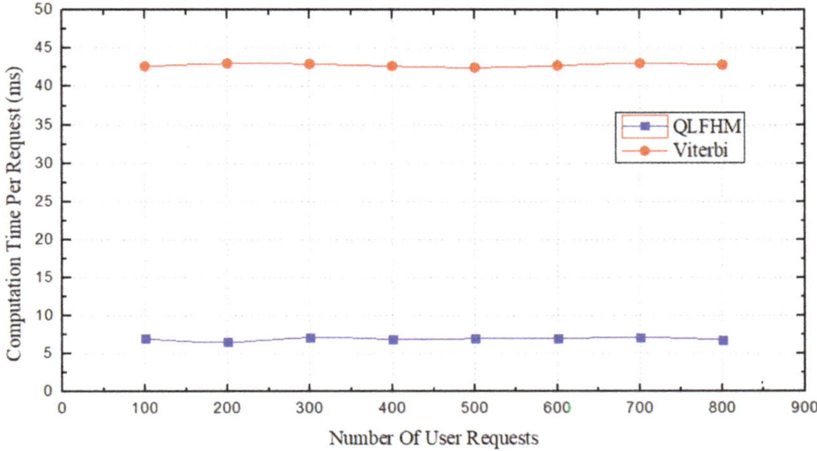

Figure 8. Comparison of computation time per request.

By comparison, we can draw the following conclusion. Under the condition that the output deployment scheme is close to the optimal solution, the Q algorithm provides a result faster than does the general heuristic algorithm. The request acceptance ratio and the service provider average profit values indicate that the algorithm considers the global network insofar as possible—that is, it better guarantees the load balance.

5.3.2. Effects of the Use Ratio λ

We know that the Q matrix stores several link strategies between any two points in the topology. However, in the actual output process, scoring all the links may not be the best option. Here, we test the proportion of the use of the RL output, which we call the use ratio λ.

There are three parameters $x1$, $x2$, $x3$ associated with λ in step 4 of Algorithm 4. The parameter $x1$ represents the ratio of the recommended value of the next action in a state. For example, if $x1$ is set to 0.6, alternative paths can be added if their recommended value is greater than 0.6 multiplied by the maximum recommended value. The parameter $x2$ limits the number of paths to find. For example, when $x2$ is set to 100, the algorithm will stop looking after finding 100 candidate paths. The parameter $x3$ represents the longest length of a single path, depending on the number of VNFs required.

To perform a comparison, we divide λ into three scenarios: low λ, balanced λ and high λ. The use ratio design is listed in Table 2.

Table 2. The design of the use ratio λ .

Use Ratio	λ_{low}	$\lambda_{balanced}$	λ_{high}
$x1$	0.79	0.6	0.4
$x2$	100	100	100
$x3$	7	7	7

After completing this analysis, we selected the balanced parameter (which is the λ parameter used in the previous section). We used the same three metrics (request acceptance ratio, service provider average profit, and computation time per request) for analysis.

From Figure 9, we can see the acceptance ratio for requests in the three λ states. When the number of requests is less than 400, the acceptance ratio of the three λ states is not particularly stable; of these, the fluctuations of λ_{high} and λ_{low} are high, while the fluctuation of $\lambda_{balanced}$ is much higher than the other two λ states. When the request number is greater than 400, the acceptance ratio of all three λ states tends to be stable, but the acceptance ratio of $\lambda_{balanced}$ is approximately 10% higher than those of the other two states. This is because when the λ state is λ_{high}, some longer paths may obtain high scores; thus, they will be selected for deployment, occupy more bandwidth resources, and affect other SFC deployments. When the λ state is λ_{low}, the number of options for participation is insufficient, and the optimal choice cannot be found.

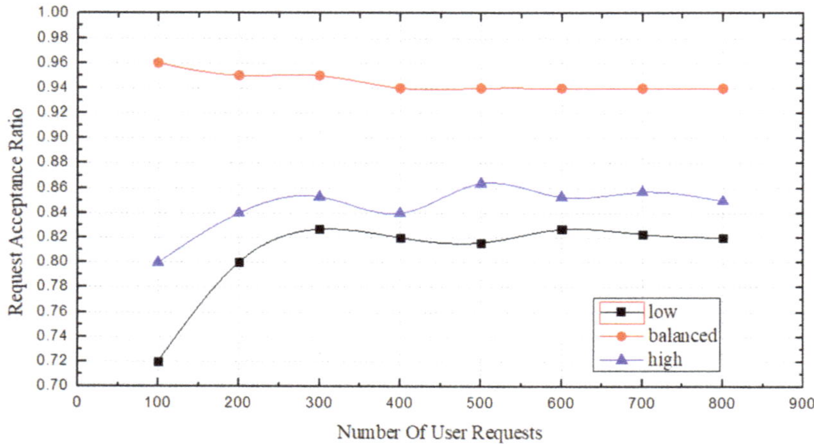

Figure 9. Comparison of the request acceptance ratio among different λ states.

In Figure 10, we compare the service provider's profits in the three λ states. The profit difference among the three states is not obvious when the number of requests is low; however, as the number of requests increases, the profit margins in the λ_{high} and λ_{low} states remain close and their slopes are approximately the same. In contrast, the profit of $\lambda_{balanced}$ increases at a greater slope, increasing the gap between its profits and those of the other λ states. This result is related to the deployment success rate and the length of the deployed SFCs. We are confident that when the λ state is $\lambda_{balanced}$, the profit obtained by using this algorithm will be larger than the profits obtainable using the other two states of λ.

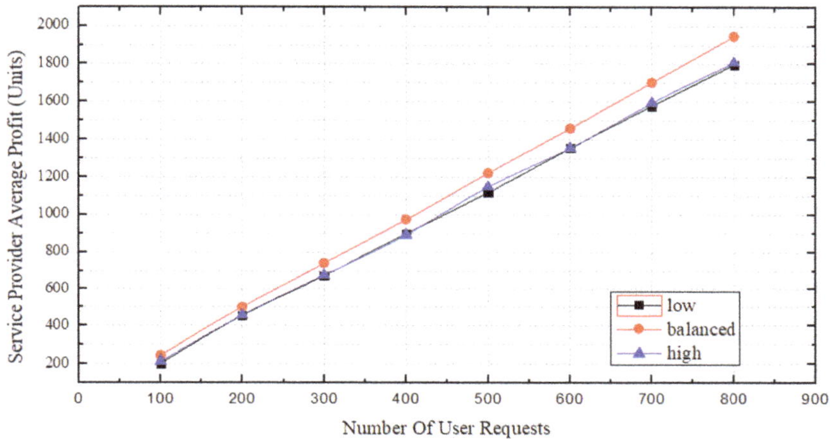

Figure 10. Comparison of service provider average profit among different λ states.

Figure 11 shows a comparison of operation times under the three λ states. The average operation time in the λ_{low} and $\lambda_{balanced}$ states is largely stable, while some fluctuation occurs in the λ_{high} state. The value of λ determines the number of candidate paths that participate in the score stage; consequently, the operation time is proportional to the value of λ. However, as shown, we found that the average operation time of the $\lambda_{balanced}$ value is closer to the λ_{low} state and better matches the desired time efficiency.

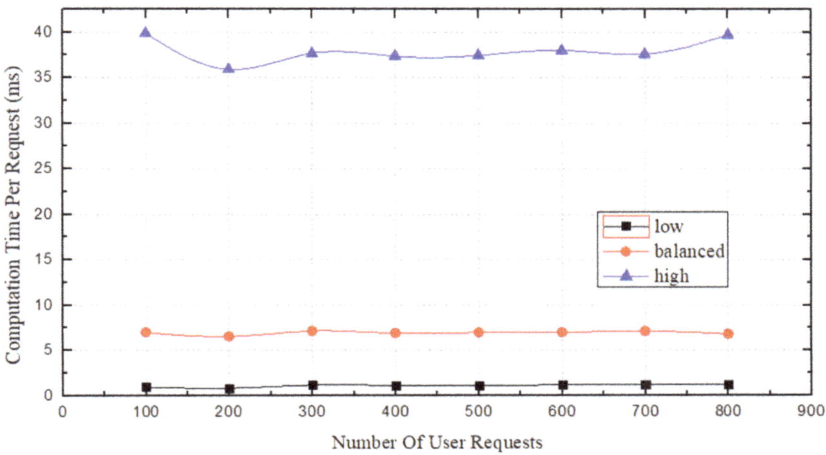

Figure 11. Comparison of computation time per request among different λ states.

After conducting this comparison, we think that the $\lambda_{balanced}$ setting is better than the λ_{high} or λ_{low} settings. These results suggest that in some cases, the local optimum is not the global optimum. On one hand, the $\lambda_{balanced}$ setting helps to reduce the output time. On the other hand, it can reduce the operational overhead of the service provider. Overall, the setting represents a tradeoff between reducing execution time and achieving an optimal solution.

5.3.3. Comparison of Training Time

In this sub-section, we show a comparison of operation time between original and optimized Q-learning training algorithm. We change h_{max} to make the path we need to find longer, and the number of paths increases, which is the same as the case when the network topology keeps getting bigger. Taking the convergence of Q matrix as the termination time, the comparison results are shown in Figure 12.

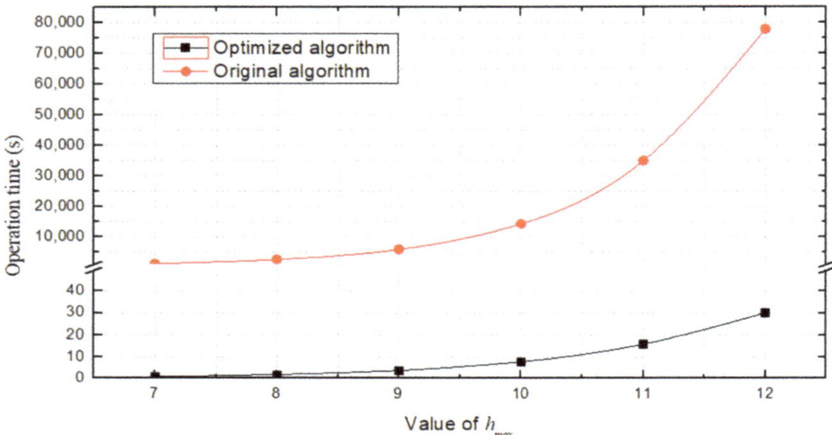

Figure 12. comparison of the operation time between Original and Optimized Q-learning Training Algorithm.

From Figure 12 we can see that the operation time is not at an order of magnitude, and the gap grows larger and larger with the increase of h_{max}. This is because the times of trial and error training iterations of the RL is very large. Take h_{max} as 7 for example, there are 10,000 iterations in 5 s, and 2,252,000 iterations are needed to make the Q matrix to converge. When h_{max} is larger, it is more difficult to achieve the convergence of Q matrix.

From the comparison results, it can be seen that in any practical situation, the optimized algorithm can produce a convergent Q matrix within 1 min. This not only solves the problem of reducing the training time, but also solves the problem of judging whether the Q matrix converges.

The entire simulation comparison shows that compared with the benchmark algorithms, the QLFHM algorithm proposed in this paper has excellent performance and can make decisions based on requests in a very short time while maximizing the service provider's profits. Moreover, the three adjustable parameters not only increase the algorithm's flexibility but also leave room for improvement.

6. Conclusions

This study investigated SFC deployment in a dynamic network. We designed an effective algorithm (QLFHM) to solve this problem. In the QLFHM algorithm, we consider the network load balance and make corresponding countermeasures to reflect the real-time changes in a dynamic network. The algorithm first reads the network topology information and then learns the topology routing scheme through the RL module. Then, it uses the load balancing module to select the optimal solution from several candidate schemes output by the RL module. The improved learning algorithm improves the efficiency of addressing this specific type of problem; it not only capitalizes on the decision-making advantages of RL but also avoids a lengthy training process. Finally, we conducted extensive simulation experiments in a simulated network environment to evaluate the performance of our proposed algorithm. The experimental results show that the performance of the proposed QLFHM algorithm is superior to that of the benchmark algorithms CG and Viterbi when processing service requests. We are confident that while QLFHM algorithm ensures the security of user data, its performance advantages are reflected by the decision time, load balancing, deployment success rate and deployment profit when deploying SFCs.

In future work, we will further carry out other related researches such as the migration of virtual machines for the deployed VNFs [38], the energy-saving operation of servers in the network [39], and the decentralization of resource allocation controllers [40] to extend our current study.

Author Contributions: Conceptualization, J.S.; Methodology, H.Y.; Software, V.C.; Writing—Original Draft Preparation, G.H.; Writing—Review & Editing, A.K.S.; Supervision & Project Administration, G.S.

Funding: This research was funded by National Natural Science Foundation of China grant number [61571098], Fundamental Research Funds for the Central Universities grant number [ZYGX2016J217], and the 111 Project grant number [B14039].

Conflicts of Interest: The authors declare no conflict of interest.

References

1. Sun, G.; Chang, V.; Guan, S. Big Data and Internet of Things—Fusion for different services and its impacts. *Future Gener. Comput. Syst.* **2018**, *86*, 1368–1370. [CrossRef]
2. Shah, S.; Wu, W.; Lu, Q. AmoebaNet: An SDN-enabled network service for big data science. *J. Netw. Comput. Appl.* **2018**, *119*, 70–82. [CrossRef]
3. Sun, G.; Anand, V.; Liao, D. Power-efficient provisioning for online virtual network requests in cloud-based data centers. *IEEE Syst. J.* **2015**, *9*, 427–441. [CrossRef]
4. Wu, K.; Lu, P.; Zhu, Z. Distributed online scheduling and routing of multicast-oriented tasks for profit-driven cloud computing. *IEEE Commun. Lett.* **2016**, *20*, 684–687. [CrossRef]
5. Sun, G.; Liao, D.; Zhao, D. Towards Provisioning Hybrid Virtual Networks in Federated Cloud Data Centers. *Future Gener. Comput. Syst.* **2018**, *87*, 457–469. [CrossRef]
6. Herrera, J.G.; Botero, J.F. Resource Allocation in NFV: A Comprehensive Survey. *IEEE Trans. Netw. Serv. Manag.* **2017**, *13*, 518–532. [CrossRef]
7. Yi, B.; Wang, X.; Li, K. A comprehensive survey of Network Function Virtualization. *Comput. Netw.* **2018**, *133*, 212–262. [CrossRef]
8. Mijumbi, R.; Serrat, J.; Gorricho, J.L. Network Function Virtualization: State-of-the-Art and Research Challenges. *IEEE Commun. Surv. Tutor.* **2016**, *18*, 236–262. [CrossRef]
9. Sun, G.; Chang, V.; Yang, G. The Cost-efficient Deployment of Replica Servers in Virtual Content Distribution Networks for Data Fusion. *Inf. Sci.* **2018**, *432*, 495–515. [CrossRef]
10. Fang, W.; Zeng, M.; Liu, X. Joint spectrum and IT resource allocation for efficient vNF service chaining in inter-datacenter elastic optical networks. *IEEE Commun. Lett.* **2016**, *20*, 1539–1542. [CrossRef]
11. Ghanwani, A.; Krishnan, R.; Kumar, N. Service Function Chaining (SFC) Operation, Administration and Maintenance (OAM) Framework. *J. Am. Chem. Soc.* **2017**, *90*, 543–552. [CrossRef]
12. Fang, W.; Lu, M.; Liu, X. Joint defragmentation of optical spectrum and IT resources in elastic optical datacenter interconnections. *J. Opt. Commun. Netw.* **2015**, *7*, 314–324. [CrossRef]
13. Moens, H.; Turck, F. Customizable function chains: Managing service chain variability in hybrid NFV networks. *IEEE Trans. Netw. Serv. Manag.* **2016**, *13*, 711–724. [CrossRef]
14. Liu, J.; Lu, W.; Zhou, F. On dynamic service function chain deployment and readjustment. *IEEE Trans. Netw. Serv. Manag.* **2017**, *14*, 543–553. [CrossRef]
15. Mars, P.; Chen, J.R. *Learning Algorithms: Theory and Applications in Signal Processing, Control and Communications*; CRC Press: Boca Raton, FL, USA, 2018.

16. Apostolopoulos, P.A.; Tsiropoulou, E.E.; Papavassiliou, S. Demand Response Management in Smart Grid Networks: A Two-Stage Game-Theoretic Learning-Based Approach. *Mob. Netw. Appl.* **2018**, 1–14. [CrossRef]
17. Tsiropoulou, E.E.; Katsinis, G.K.; Filios, A. On the Problem of Optimal Cell Selection and Uplink Power Control in Open Access Multi-service Two-Tier Femtocell Networks. In Proceedings of the International Conference on Ad-Hoc Networks and Wireless, Benidorm, Spain, 22–27 June 2014; pp. 114–127.
18. Xiong, R.; Cao, J.Y.; Yu, Q.Q. Reinforcement learning-based real-time power management for hybrid energy storage system in the plug-in hybrid electric vehicle. *Appl. Energy* **2018**, *211*, 538–548. [CrossRef]
19. Radac, M.B.; Precup, R.E. Data-driven model-free slip control of anti-lock braking systems using reinforcement Q-learning. *Neurocomputing* **2018**, *275*, 314–329. [CrossRef]
20. Xiao, L.; Lu, X.; Xu, D. UAV Relay in VANETs Against Smart Jamming with Reinforcement Learning. *IEEE Trans. Veh. Technol.* **2018**, *67*, 4087–4097. [CrossRef]
21. Unsal, C.; Kachroo, P.; Bay, J.S. Multiple stochastic learning automata for vehicle path control in an automated highway system. *IEEE Trans. Syst. Man Cybern. Part A Syst. Hum.* **2002**, *29*, 120–128. [CrossRef]
22. Barto, A.G.; Anandan, P.; Anderson, C.W. Cooperativity in networks of pattern recognizing stochastic learning automata. In *Adaptive and Learning Systems*; Springer: Boston, MA, USA, 1986; pp. 235–246.
23. Khazaei, M. Occupancy overload control by Q-learning. *Lect. Notes Electr. Eng.* **2019**, *480*, 765–776. [CrossRef]
24. Kai, A.; Deisenroth, M.P.; Brundage, M. Deep Reinforcement Learning A brief survey. *IEEE Signal Process. Mag.* **2017**, *34*, 26–38. [CrossRef]
25. Seeliger, K.; Güçlü, U.; Ambrogioni, L. Generative adversarial networks for reconstructing natural images from brain activity. *Neuroimage* **2018**, *181*, 775–785. [CrossRef] [PubMed]
26. Sun, G.; Liao, D.; Bu, S. The Efficient Framework and Algorithm for Provisioning Evolving VDC in Federated Data Centers. *Future Gener. Comput. Syst.* **2017**, *73*, 79–89. [CrossRef]
27. Sun, G.; Liao, D.; Anand, V. A New Technique for Efficient Live Migration of Multiple Virtual Machines. *Future Gener. Comput. Syst.* **2016**, *55*, 74–86. [CrossRef]
28. Bari, M.F.; Chowdhury, S.R.; Ahmed, R. Orchestrating virtualized network functions. *IEEE Trans. Netw. Serv. Manag.* **2016**, *13*, 725–739. [CrossRef]
29. Li, D.; Lan, J.L.; Wang, P. Joint service function chain deploying and path selection for bandwidth saving and VNF reuse. *Int. J. Commun. Syst.* **2018**, *31*. [CrossRef]
30. Sun, G.; Liao, D.; Zhao, D. Live Migration for Multiple Correlated Virtual Machines in Cloud-based Data Centers. *IEEE Trans. Serv. Comput.* **2018**, *11*, 279–291. [CrossRef]
31. Luizelli, M.C.; Bays, L.R.; Buriol, L.S. Piecing together the NFV provisioning puzzle: Efficient placement and chaining of virtual network functions. In Proceedings of the IFIP/IEEE International Symposium on Integrated Network Management (IM), Ottawa, ON, Canada, 11–15 May 2015; pp. 98–106.
32. Sun, G.; Li, Y.; Liao, D. Service Function Chain Orchestration across Multiple domains: A Full Mesh Aggregation Approach. *IEEE Trans. Netw. Serv. Manag.* **2018**, *15*, 1175–1191. [CrossRef]
33. Gupta, A.; Habib, M.F.; Chowdhury, P. *Joint Virtual Network Function Placement and Routing of Traffic in Operator Network*; Technical Report; University of California Davis: Davis, CA, USA, 2015.
34. Sun, G.; Li, Y.; Yu, H. Energy-efficient and Traffic-aware Service Function Chaining Orchestration in Multi-Domain Networks. *Future Gener. Comput. Syst.* **2019**, *91*, 347–360. [CrossRef]
35. Sun, G.; Li, Y.; Li, Y. Low-Latency Orchestration for Workflow-Oriented Service Function Chain in Edge Computing. *Future Gener. Comput. Syst.* **2018**, *85*, 116–128. [CrossRef]
36. Kim, S.I.; Kim, H.S. A research on dynamic service function chaining based on reinforcement learning using resource usage. In Proceedings of the International Conference on Ubiquitous & Future Networks, Milan, Italy, 4–7 July 2017; pp. 582–586.
37. Tchana, A.; Tran, G.S.; Broto, L. Two levels autonomic resource management in virtualized IaaS. *Future Gener. Comput. Syst.* **2013**, *29*, 1319–1332. [CrossRef]
38. Teabe, B.; Tchana, A.; Hagimont, D. Enforcing CPU allocation in a heterogeneous IaaS. *Future Gener. Comput. Syst.* **2015**, *53*, 1–12. [CrossRef]

39. Tchana, A.; Palma, N.D.; Safieddine, I. Software consolidation as an efficient energy and cost saving solution. *Future Gener. Comput. Syst.* **2016**, *58*, 1–12. [CrossRef]
40. Gueye, S.M.K.; de Palma, N.; Rutten, É. Coordinating self-sizing and self-repair managers for multi-tier systems. *Future Gener. Comput.Syst.* **2014**, *35*, 14–26. [CrossRef]

Article

A New Meta-Heuristic Algorithm for Solving the Flexible Dynamic Job-Shop Problem with Parallel Machines

Arun Kumar Sangaiah [1], Mohsen Yaghoubi Suraki [2], Mehdi Sadeghilalimi [3], Seyed Mostafa Bozorgi [4], Ali Asghar Rahmani Hosseinabadi [5] and Jin Wang [6,*]

1 School of Computing Science and Engineering, Vellore Institute of Technology (VIT), Vellore 632014, India; arunkumarsangaiah@gmail.com
2 Department of IT and Computer Engineering Qazvin Branch, Islamic Azad University, Qazvin 1519534199, Iran; M.Yaghoubi@qiau.ac.ir
3 Department Hamun Islamic Azad University of Qazvin 1519534199, Iran; Imehdisadeghi@gmail.com
4 Department of Computer Engineering, Tehran North Branch, Islamic Azad University, Tehran 1651153311, Iran; S.M.Bozorgi@iau-tnb.ac.ir
5 Young Researchers and Elite Club, Ayatollah Amoli Branch, Islamic Azad University, Amol 4865116915, Iran; A.R.Hosseinabadi@iaubeh.ac.ir
6 School of Computer & Communication Engineering, Changsha University of Science & Technology, Changsha 410004, China
* Correspondence: jinwang@csust.edu.cn; Tel.: +86-180-148-49250

Received: 17 December 2018; Accepted: 28 January 2019; Published: 1 February 2019

Abstract: In a real manufacturing environment, the set of tasks that should be scheduled is changing over the time, which means that scheduling problems are dynamic. Also, in order to adapt the manufacturing systems with fluctuations, such as machine failure and create bottleneck machines, various flexibilities are considered in this system. For the first time, in this research, we consider the operational flexibility and flexibility due to Parallel Machines (PM) with non-uniform speed in Dynamic Job Shop (DJS) and in the field of Flexible Dynamic Job-Shop with Parallel Machines (FDJSPM) model. After modeling the problem, an algorithm based on the principles of Genetic Algorithm (GA) with dynamic two-dimensional chromosomes is proposed. The results of proposed algorithm and comparison with meta-heuristic data in the literature indicate the improvement of solutions by 1.34 percent for different dimensions of the problem.

Keywords: Dynamic job-shop; Parallel Machines; Maximum flow-time of components; Genetic Algorithm

1. Introduction

Scheduling (assigning resources to tasks) is one of the most important decisions in the optimal utilization of facilities and equipment. Research in this area can be divided into two parts—static environments and dynamic environments. Most of the methods discussed in the literature can solve the static scheduling but do not consider dynamic events, such as different times of new orders, staff reductions, and machine failure. In the scheduling problems, if the set of tasks don't change over time, then the manufacturing system is static. In contrast, if new a task is added to the set of tasks during the time, this system is a dynamic manufacturing system [1,2]. In other words, in a dynamic manufacturing system, all tasks are available (at zero time) simultaneously. But, in an actual Sequence of operations, the runtime of tasks needs to be considered. In practice, scheduling problems are dynamic and the main reason for that is that some of the mentioned parameters change over the time. Therefore, the dynamic nature of a scheduling problem should be considered to solve practical problems [3,4]. It has been proved that solving static models is easier than dynamic models and wide reviews have been done on them [5].

Current research, consider the flexibility of operation and flexibility due to parallel machines in addition to considering the dynamic manufacturing environment (due to the non-zero entry time of tasks to the workshop).

In general, in classic manufacturing models, there is only one processing route for each task, which leads to the lack of flexibility in such production systems. In such environments, production flexibility is a key competitive weapon, and having this capability becomes a competitive advantage that causes to able them to quickly respond to unpredictable changes.

For example, having some flexibility in an industrial production context for powerful manufacturing systems leads to being adaptable to fluctuations, such as machine failure and bottlenecking of machine creation. And then each machine can produce several products [6,7].

Today, most manufacturing systems keep several versions of each machine at each workstation (flexibility due to parallel machines) to solve the bottleneck problem (due to long processing times for some components or due to machine failure) and to increase production and improve performance. Some operations can be processed not only on one station but also on a set of stations available in the workshop (flexibility due to operation). These conditions add another complexity to this problem, so finding an approximately optimal solution for these problems is very complicated and difficult [8–12].

In the literature, these problems are known as flexible scheduling problems. In a comprehensive survey that was conducted in 2000, the dimensions of this flexibility were expanded to 15 [13].

In this paper, Flexible Dynamic Job-Shop with Parallel Machines (FDJSPM) in manufacturing environments has been studied. There are n components and m workstations in which each component needs a certain set of operations (O) and the sequence of movement between workstations is different for each component. In this study, we consider a dynamic workshop in order to be more relevant for actual manufacturing environments. Each component enters the workshop at time r (a random and non-zero time). To avoid bottlenecks and increase production rates, each workstation can process more than one component (due to operation flexibility). Each workstation M includes L parallel machines with non-uniform speeds and can process the assigned tasks (because of the flexibility of parallel machines with uniform speeds).

The aim of this paper is to model the problem and then solve the problem of minimizing the workflow (F_{max}) by considering the dynamics generated in the workshop, i.e., operation flexibility and the flexibility of the parallel machines with non-uniform speed. In this study, for the first time, we consider the operational flexibility and the flexibility due to parallel machines with non-uniform speed for a dynamic job shop.

Accordingly, the FDJSPM problem splits into two subproblems: allocation and sequencing of operations. The allocation subproblem is caused by the flexibility generated in job-shop production, and increases the complexity of the problem.

This paper consists of 7 sections. The introduction and research necessity are investigated in this section. The literature on the subject and related work in the field of flexible, dynamic, and static manufacturing systems are presented in Section 2. The problem definition and the mathematical model are introduced in Section 3. The proposed approach is defined in Section 4. Numerical experiments and experimental results are presented in Sections 5 and 6, respectively. Results of the research and future work are presented in Section 7. The following, literature review and research background in the field of "job-shop", "flexible job-shop" and "scheduling by parallel machines with uniform and non-uniform speeds" are presented.

2. Related Work

Classic Job-Shop Scheduling (JS) and Dynamic Job-Shop Scheduling (DJS) are the most important production management problems and one of the most difficult combinatorial optimization problems.

Generally, research done on scheduling in dynamic environments has been based on either queuing theory or rolling time horizon. In the first case, the scheduling problem is considered as a queuing system and new orders go on devices after arrival and they are processed based on their prioritization.

Analytical methods for solving this problem are based on queuing theory [1]. Due to the complexity of the problem, the above-mentioned analytical methods can solve the problems with one machine. The rolling time horizon technique is an efficient way to provide schedules for dynamic problems in which the dynamic scheduling problem is split into several static scheduling problems and in each stage, each problem is solved statically. This method was applied to solve medium-term scheduling problems [1]. Then, it was used to investigate the scheduling problem and significant results were presented in this field [14]. Later, researchers used general solutions to use this method in backward scheduling [15]. Recently, scheduling problems in dynamic and uncertain environments have been investigated by using RTH methods and their performances evaluated in different situations.

In the JS problem, the route of tasks is fixed and it isn't necessary to have the same route for all tasks. In this problem, we assume that there is one processing route for each task, which leads to the lack of flexibility in manufacturing systems.

Flexible Job Shop Scheduling (FJS) is an extension of JS, in which each operation can be processed by a set of machines. According to the definition, FJS includes two subproblems: routing (assign the machine to the operation) and scheduling (sequencing the operations) [16].

FJSPM is an extension of FJS and parallel machines. In each step of FJS, there is only one machine. In FJSPM, there is more than one machine at least in one step, and in each step, a set of parallel machines are placed together and each of them can process the operation assigned to that step. Thus, we can choose a different route to process the operations. In fact, a new kind of flexibility is defined for JS which has not been considered in the literature. The main idea of this flexibility is to increase the production, address the bottleneck problem, and use it as a competitive advantage in the economic environment.

In 1998 [17], mixed three-stage JS with a particular structure; one machine in step 1 and 3 and two machines in step 2, was investigated with the aim of minimizing the maximum completion time (C_{max}). In the same year, a Taboo Search Algorithm (TS) was also presented for FJSPM in order to reduce the C_{max} [18].

In 1999 [19] a study proposed a multi-objective algorithm to solve the flexible process. They formulated the machine loading problem as a two-criteria integer program and proposed two different meta-heuristic models, one of which is based on theory and the other is based on the GA. In the same year, a hybrid genetic-based algorithm was proposed to minimize the C_{max} in FJS [20]. After that, in 2001, a polynomial algorithm with/without considering interruption was proposed for FJS [21]. In 2002, for the first time, FJS was investigated in a multi-purpose case [22], and the proposed algorithm was a hybrid of fuzzy logic and evolutionary algorithms. The flexibility of the problem was limited only to the operational flexibility. Fuzzy logic is used to search the target space in order to find the Pareto solutions.

In the same year, a GA was proposed for the same FJS problem in the supply chain to reduce the C_{max} [23]. They assume the existence of external sources, machine replacement for each operation, and sequences of multiple operations for each component.

In 2003 [24], a symbiotic evolutionary algorithm (an evolutionary algorithm) which is based on artificial intelligence was proposed. Some researchers tried to solve the FJS problem by using heuristic methods and fuzzy logic [25]. In 2004, the flexibility in JS was limited to the operation flexibility [26]. The aim of this research was to reduce the latency and TS was used to solve the problem. In 2004, a GA was proposed for the problem that researchers had discussed in 2002 [27]. In this study, at first, the allocation subproblem is solved by the priority rule SPT and then, based on the most important step of finding the best answer (selecting an appropriate representation of the chromosomes step), a GA with chromosome representation based on the base operations is proposed. Computational results show the effectiveness of the algorithm in dealing with large problems.

The simulated annealing algorithm (SA) is proposed for Multi-Stage Job Shop with Parallel Machines (MSJSPM) with the aim of minimizing the total current time [28]. Also, problems studied in previous years [22] were investigated in this year [29] and a hybrid hierarchy method was proposed for that problem. In this method, Particle Swarm Optimization (PSO) is used for solving the routing subproblem, and SA is used for solving the sequence of operations. In 2004, a new method was proposed for economic scheduling of multi-products with a general cycle in FJS [30] and a multi-objective and efficient practical hierarchy method was proposed for FJS [31]. This method is a hybrid of PSO and SA. Then, a new algorithm based on the Artificial Immune System (AIS) for FJS with turning back was proposed [32].

In 2006, FJS with uniform parallel machines, with the aim of minimizing the C_{max}, was studied [33]. They developed a set of meta-heuristic algorithms and showed that for a large number of tasks, a vector addition based meta-heuristic is optimal. Researchers used three kinds of flexibility; totally (FJSP-100), average (FJSP-50), and partial (FJSP-20) [3]. They said that C% flexibility means all operations of different tasks can be processed by C% of available machines in JS [1]. In 2007, overreliance of evolutionary algorithms on recombination mechanisms and random selection was the most important limitation of evolutionary algorithms [34].

In order to overcome these limitations and obtain efficient solutions for FJSP, they tried to take advantage of the interaction between learning and evolution, and then they presented their GA, named Hybrid Genetic Algorithm (HGA). Their algorithm was a hybrid of the evolutionary algorithm, learning pattern, and population generation. A simple distribution rule used for population generation and K-nearest neighborhood were used for the learning pattern. A combination of the Bottleneck transmission method and a GA was used for FJS with three objectives in 2007 [35]. They utilized the chromosome display method in [27] which uses 2 vectors instead of 1 vector for machine assignment and sequence of operations. Also, they utilized the global search capability of GA and the local search capability of the Bottleneck transmission method to solve the problem. In [30,31], the authors examined scheduling in flexible flow lines with sequence-dependent setup times to minimize the makespan. This type of manufacturing environment is found in industries, such as printed circuit board and automobile manufacture. An integer program that incorporates these aspects of the problem is formulated and discussed. Because of the difficulty in solving the IP directly, several heuristics are developed, based on greedy methods, flow line methods, the Insertion Heuristic for the Traveling Salesman Problem (TSP), and the Random Keys GA.

Generally, the results of a comprehensive study on the fields of "job shop", "flexible job shop", and "job-shop scheduling with uniform/non-uniform parallel machine", show that in most sequences, all tasks are assumed to be available simultaneously. But, in real manufacturing environments, the set of tasks which must be scheduled vary over time. This fact shows the dynamicity of the job-shop problem. In addition, being adaptable to fluctuations, such as machine failure and bottlenecks of machine creation, is very important for such systems.

These adaptations are generated by considering base flexibilities. In such a way, manufacturing systems could use them as competitive advantages in addition to overcoming such fluctuations. Therefore, by reviewing the literature and identify existing gaps, we investigate the FDJSPM.

For the first time, in this research, we consider "operational flexibility" and "flexibility due to parallel machines with non-uniform speed" in the field of the dynamic JS model. In the following, we model the problem with the aim of minimizing the flow time (F_{max}).

3. Problem Definition and Mathematical Model

In this section, we will present the variables and introduce the FDJSPM problem, then present an integer programming model, and finally prove the NP-hard feature of the problem.

3.1. Problem Definition

The FDJSPM problem is defined as follow: there are m processing steps (workstations) for n tasks (component) in JS where each task needs a set of operations. The Ith component enters the dynamic job-shop at a non-zero time (r_i). The Jth component consists of a chain of operations, $\{o_i, 1, o_i, 2, \ldots, o_i, n_i\}$ which, without losing the generality, we can assume is the same as the order of processing tasks in JS. The order of movement between workstations is different for various tasks (the main feature of manufacturing tasks). The number of operations needed for task completion is smaller or equal to the number of processing steps. In each step of processing, M_k, we need l_k versions of parallel machines with various speeds. Each of them can process the assigned operations to that step. In a flexible workstation, this parameter (l_k) is larger than 1. Each operation can be processed by at least one workstation, and because of the operational flexibility, there is at least one operation which can be processed on more than on a workstation.

The time needed to process the operation $O_{i,j}$ on a machine with $S_{k,pm} = 1$ in the kth step is equal to $P_{pm,j,k}$, if this operation is processed on a machine with $S_{k,pm} > 1$ the time may be reduced to $P_{i,j,k}/S_{k,pm}$. The aim of this paper is to reduce the maximum flow time of components (F_{max}).

According to the definition, the FDJSPM problem is decomposed into two sub-problems: assignment subproblem (assign tasks to flexible workstations) and sequencing operations (Sequencing operations among the tasks in the waiting queue of each workstation). The assignment subproblem is further decomposed into two subproblems: workstation assignment and machine assignment. The assignment subproblem is caused by the operational flexibility and parallel machines in JS which increase the complexity of the problem. The aim of this paper is to minimize the F_{max} by considering the following assumptions:

1. At a given time, each machine can process only one operation.
2. JS is dynamic and tasks enter to JS at a non-zero time.
3. At a given time, each task can be processed by only on one machine.
4. All machines are available at time 0 and never break down.
5. Interrupting the operation is not allowed and the processing time of each task is definite and known.
6. Available depot is allowed and its capacity is infinite.
7. Preparation time between operations is negligible. It's either a part of the processing time or the transportation time and can be ignored.

Some of the above assumptions are unrealistic in practice and are considered for simplifying the model.

3.2. Sets, Parameters, and Variables Definition

n: The number of entered tasks to the JS at the non-zero time
n_i: The number of operations of ith component
r_i: Entered the time of ith component to the JS
m: The number of processing steps (workstations)
k: Index of steps (workstations) $k = 1, \ldots, m$
M_k: k^{th} step of processing
$M_{k,r}$: r^{th} parallel machine of kth step
$O_{i,j}$: i^{th} operation of jth task
$St_{i,j}$: Step in which operation $o_{i,j}$ will be processed
l_k: The number of parallel machines in k^{th} step
l_{max}: Maximum of l_k
pm: Index of parallel machines in workstations; pm=1, …, l_k

$S_{k,r}$: Speed of $M_{k,r}$
$P_{i,j}$: Time units to process the operation $o_{i,j}$ on a machine with unit speed
A: An optional big number
F_{max}: Maximum flow time
$c_{i,j}$: Earliest finish time of $o_{i,j}$
$Ft_{k,pm}^{(i,j)}$: Finish time of $o_{i,j}$ on machine $M_{k,pm}$

$a_k^{(i,j)} = \begin{cases} 1 \\ 0 \end{cases}$ If machines in k^{th} step are possible machines for $o_{i,j}$, this variable is 1, otherwise, it is 0.

$X_{k,pm}^{(i,j)} = \begin{cases} 1 \\ 0 \end{cases}$ If $o_{i,j}$ be processed on machine $M_{k,pm}$ this variable is 1, otherwise, it is 0.

$R_{k,pm}^{(i,j)(p,q)} = \begin{cases} 1 \\ 0 \end{cases}$ If $o_{p,q}$ be processed before $o_{i,j}$ on $M_{k,pm}$ this variable is 1, otherwise it is 0.

3.3. Mathematical Models

3.3.1. Objective Function

Minimization of flow time

$$F = \{Max\{C_i | i = 1, \ldots, N\} - r_i\} \tag{1}$$

3.3.2. Constraints of the Problem

$$\begin{gathered} Ft_{k,pm}^{(i,j+1)} - Ft_{k,pm'}^{i,j} + L \times (1 - a_k^{(i,j+1)} \times X_{k,pm}^{(i,j+1)}) \geq P_{i,j+1,k} / S_{k,pm} \\ 1 \leq i \leq n; 1 \leq j \leq (n_i - 1);\ 1 \leq k, k' \leq m;\ 1 \leq pm, pm' \leq l_{k'} \end{gathered} \tag{2}$$

$$\begin{gathered} Ft_{k,pm}^{(i,j)} - Ft_{k,pm'}^{(p,q)} \times X_{k,pm}^{(i,j)} + L \times R_{k,pm}^{(i,j)(p,q)} \geq X_{k,pm}^{(i,j)} \times P_{i,j,k} / S_{k,pm} \\ i = 1, \ldots, n-1;\ 1 \leq j \leq n_i;\ q = i+1, \ldots, n;\ 1 \leq p \leq n_q;\ 1 \leq pm, pm' \leq l_{k'} \end{gathered} \tag{3}$$

$$\begin{gathered} Ft_{k,pm}^{(p,q)} - Ft_{k,pm}^{(i,j)} \times X_{k,pm}^{(p,q)} + L \times (1 - R_{k,pm}^{(i,j)(p,q)}) \geq X_{k,pm}^{(p,q)} \times P_{p,q,k} / S_{k,pm} \\ i = 1, \ldots, n-1;\ 1 \leq j \leq n_i;\ 1 \leq k \leq m;\ 1 \leq pm \leq l_{k'};\ q = i+1, \ldots, n;\ 1 \leq p \leq n_q \end{gathered} \tag{4}$$

$$\sum_{k=1}^{m} \sum_{pm=1}^{l_k} a_k^{(i,j)} \times X_{k,pm}^{(i,j)} = 1 \qquad 1 \leq i \leq n; 1 \leq j \leq n_i \tag{5}$$

$$X_{k,pm}^{(i,j)} \leq a_k^{(i,j)} \qquad 1 \leq i \leq n; 1 \leq j \leq n_i;\ 1 \leq k \leq m;\ 1 \leq pm \leq l_{k'} \tag{6}$$

$$Ft_{k,pm}^{(i,1)} \leq A\, X_{k,pm}^{(i,1)} \qquad 1 \leq i \leq n;\ 1 \leq k \leq m;\ 1 \leq pm \leq l_{k'} \tag{7}$$

$$C_i \geq \sum_{k=1}^{m} \sum_{pm=1}^{l_k} Ft_{k,pm}^{(i,n_i)} \qquad 1 \leq i \leq n; \tag{8}$$

$$Ft_{k,pm}^{(i,j)} \leq L \times X_{k,pm}^{(i,j)} \times P_{i,1,k} / S_{k,pm} + r_i\ ; 1 \leq i \leq n; \forall i; 1 \leq j \leq n_i; 1 \leq k \leq m; 1 \leq pm \leq l_{k'} \tag{9}$$

3.4. Model Description

In order for scheduling to be feasible, it's necessary to assign each operation to the right recourses [1]. Equation (2) ensures that operations don't have any interference. In other words, these Equations lead to process operations in an appropriate order.

Since two operations cannot be processed on one machine at the same time, a valid schedule must consider resource limitations to solve the problem [1]. Equations (3) and (4) ensure that operations on one processor don't have interference.

All models presented for JS have at least 2 similar limitations. All JS problems have constraints 1, 2, and 3 in common. The innovation of this paper is as follows:

(a) Equations (5) and (6) ensure each operation of each task is assigned to one machine. In other words, Equation (5) ensures that each task must be processed only on one machine and be processed just one time. These equations are generated due to the flexibility of parallel machines in workstations. Beside these equations, Equation (7) explain that if in step K, $O_{i,j}$ isn't assigned to each machine, its completion time on all machines must be 0. These equations are generated due to the flexibility considered in JS.

(b) On the other hand, because the objective function (F_{max}) is based on the completion (finish) time, Equation (8) computes the maximum finish time. Equation (9) is considered to study the dynamic property of the problem due to differences in the arrival time of jobs in the workshop.

We used lingo software to evaluate the proposed model and we solved the small and average FDJSPM problem with this model. The results confirm the validity of the proposed model to solve the FDJSPM problem.

3.5. NP-Hard Problem

The complexity of a combinatorial problem is the amount of time spent to solve that problem. The FDJSPM problem is an extension of the FDJSP problem but with flexibility due to parallel machines in a dynamic manufacturing environment. So, its complexity and hardness are the same as for the FDJSP problem. If operational flexibility is considered in these problems, the complexity of finding the optimal approximate solutions is increased [1]. Since FDJSP with operational flexibility is strongly NP-hard [8], this problem, with respect to parallel machines in a dynamic manufacturing environment, will also be strongly NP-hard.

3.6. Problem-Solving Approach

For solving such scheduling problems, we need an efficient method. This method must be able to consider the complexities arising from the exponential increase in the solution space to find a proper solution.

In this paper, we utilize the capability of a GA to design an efficient method to solve the problem. In the following section, the structure of the proposed GA for solving the FDJSPM problem is introduced and then parameters and the efficiency of the proposed algorithm based on numerical experiments are shown.

4. The Proposed Algorithm

Since the FDJSPM problem is categorized as a NP-hard problem, there is no algorithm to find an efficient solution in a reasonable time for large or medium size problems. If an algorithm is able to find a proper solution, it needs a long time. In this section, a GA is proposed which is able to find a solution in a reasonable time. Figure 1 shows the general procedure of the proposed GA.

The main components of GA are: chromosome representation, initial population generation, fitness function, crossover and mutation operator, and selection method. The parameters of the GA, including the population size, number of generations, probability of crossover (p_c), and probability of mutation (p_m), must be predefined [22].

4.1. Chromosome Representation (Problem Encoding)

The first step in JS or any optimization problem using GA is to represent the solutions in the form of chromosomes [36].

One of the capabilities of using the GA is to design the solutions corresponding to chromosomes (schedules). If chromosome design and classic operations are applied efficiently, one can avoid generating inefficient chromosomes, and it is not necessary to use the penalty or repairing strategy. In the literature on the use of genetic algorithms to solve problems, many criteria have been defined to design chromosomes. Among them, criteria such as least amount of space and time due to the computational complexity of the problem also avoid generating inefficient chromosomes [1].

The FDJSPM problem is decomposed to two subproblems: "allocation" and "sequencing the operations". The proposed GA is designed in such a way that it is able to solve these two sub-problems seamlessly and simultaneously.

For this purpose, a two-dimensional dynamic chromosome with respect to the dynamicity of entering tasks to the system is presented (Figure 1). In this representation, chromosome length is equal to all the operations for scheduling ($nTOP$) and its width is equal to 3. So, each solution is presented as a two-dimensional array.

This method is similar to the method used in 2002, in which the allocation sub-problem was decomposed into two distinct sub-problems: workstation allocation and machine allocation.

In this representation, the first row of the chromosome shows the workstation processing the desired operation and the second row is the number of the machine that will process this operation. The third row contains the priority assigned to each operation. Each element in the third row is always a number between 1 and $nTOP$ and the priority of two chromosomes cannot be the same. Thus, the obtained solution is always justified. In general, we can say that the first two rows indicate the allocation and the third row indicates the sequence of operations.

In order to understand the encoding system, consider an example containing 4 jobs and 3 workstations. A flexible manufacturing system with parallel machines is designed in such a way that each job has 3 operations and each workstation has 3, 2, 2 uniform parallel machines, respectively. Also, the manufacturing system is completely flexible. The chromosome corresponding to the solution contains 3 rows of length 12 and is encoded as in Figure 1. Chromosome representation is as follow:

The sequence is based on field operations, and the processing operation is the first of a string assigned to Station.

According to Figures 1 and 2, it is clear that each chromosome is logically divided into several parts. The number of segments of each chromosome is equal to the number of jobs. Plans for subdivisions of each job are in the same range and in the first row. In the next row of chromosomes, the overall arrangement of chromosomes is determined.

In this way, various mutation and crossover operators can be designed. Separating different aspects of the scheduling will make searching in solution space easier.

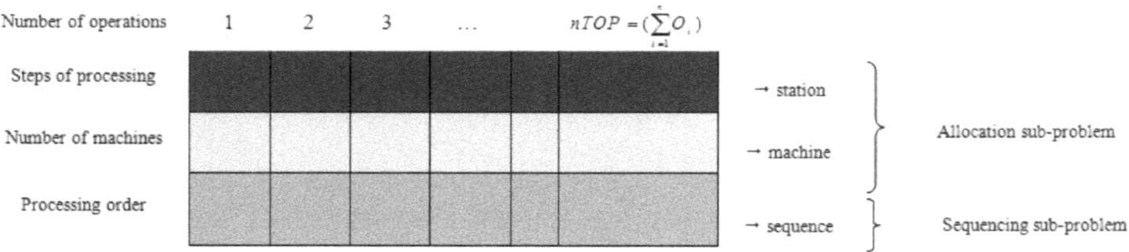

Figure 1. Chromosome representation in the proposed genetic algorithm (GA).

1	2	3	4	5	6	7	8	9	10	11	12	Location
11 O	12 O	13 O	21 O	22 O	23 O	31 O	32 O	33 O	41 O	42 O	43 O	Operation index
M_1	M_2	M_3	M_1	M_3	M_2	M_2	M_3	M_1	M_2	M_1	M_2	Workstation allocation
3	2	1	2	1	1	1	2	3	2	3	2	Machine allocation
1	8	10	5	6	9	2	3	12	4	7	11	Operation sequencing

Figure 2. Chromosome representation according to the obtained solution.

4.2. Initial Population

After determining the method of converting solutions to chromosomes, an initial chromosome must be produced. How to produce the initial population is a key step in determining the quality of the solutions in any local search method, such as a GA. Random generation of the initial population results in retention of the diversity of chromosomes in the initial population, reduces the possibility of premature convergence, and falls in local optima [35].

In the proposed GA, in order to avoid premature convergence, no heuristic method is used to produce the initial population. Thus, rows 1 and 2 are filled out by allocating a flexible workstation randomly and row 3 is filled out with a random combination of numbers 1 to $nTOP$.

4.3. Crossover Operation

The proposed GA uses two positions based on crossover and RMX crossover with the rate of P_c. Position-based crossover can be applied to the machine allocation row (rows 1 and 2 of the chromosome) and RMX crossover can be applied to the sequence of operations (row 3 of the chromosome). In position-based crossover, some genes are selected randomly from the first parent (P_1) and are exactly copied in the child chromosome (O_1). To complete the remaining genes, we use the second parent (P_2) and copy the non-repetitive genes to the same positions in O_1.

RMX crossover is applied on the sequence of operations (row 3 of chromosomes). This method was proposed by Goldberg and Langley. In fact, it is the same as 2-point crossover in the literature which is extended for special cases. In this method, two chromosomes P_1 and P_2 must be cut at a random position. This position is called the "mapping point". Genes between these two mapping points are exchanged between P_1 and P_2. Now, to create the third row of O_1, we must permutate the genes of P_1 and P_2 in O_1 and O_2 until there is no inconsistency in terms of non-identical sequence numbers. These genes are obtained by one-to-one correspondence obtained from the exchange of genes between the two cut points.

4.4. Mutation Operation

In general, mutation not only creates unplanned random changes, but also provides a wider search space and can prevent premature convergence. For this reason, in the proposed algorithm, the permutation operation is applied to each parent chromosome separately after applying the crossover operation. In this manner, for each element of a string, the possibility of mutation is tested. If this test is successful, that chromosome will be mutated.

In the proposed GA, because of the special structure of the problem and the proposed chromosome, a heuristic mutation operation based on linear inversion (a heuristic rule) is proposed. This operation works with two different mutation rates. So, for allocation, the sub-problem uses heuristic mutation with rate Pm_1, and for the sequencing operation, the sub-problem uses linear inversion with rate Pm_2. In this manner, if the testing result on a chromosome is positive, a gene from that chromosome is selected randomly and its workstation allocation field is changed to another number, which represents a flexible workstation.

To better understand the mutation operator, an example is given in this section. The chromosome selected for mutation is the chromosome of Figure 2. In this chromosome, the machine allocation and operational sequencing rows are mutated. It is assumed that O_{32} is mutated on the basis of the probability of mutation Pm_1 and further, based on the probability of the mutation Pm_2, the order of execution Job_3 is completely changed. The result of the jump operator is shown in Figure 3. The mutated sections are underlined.

1	2	3	4	5	6	7	8	9	10	11	12	Location
11	12	13	21	22	23	31	32	33	41	42	43	Operation index
O	O	O	O	O	O	O	O	O	O	O	O	
M1	M2	M3	M1	M3	M2	M2	M3	M1	M2	M1	M2	Workstation allocation
3	2	1	2	1	1	1	<u>3</u>	3	2	3	2	Machine allocation
1	8	10	5	6	9	<u>3</u>	<u>12</u>	<u>2</u>	4	7	11	Operation sequencing

Figure 3. Mutation example.

4.5. Selection Method

The proposed GA uses developed sampling space. This space is defined as follow: if the initial population size is α and after applying crossover and mutation operations β children are produced, the sampling spec size will be $\alpha + \beta$. The sampling method to produce the next generation will be elitism and the rolled-wheel selection method.

We use the elitism method to produce good solutions and increase the chance of achieving the optimal solution [28]. Based on the elitism method, $\mu + Pop - Size$ a number of chromosomes are selected and transferred to the next generation without any preconditions.

If the values of consecutive chromosomes are equal, these chromosomes aren't selected, and the next chromosomes will be tested. In GA, the λ numbers of the best chromosomes of each generation are placed directly in the next generation. The aims of this method are:

(1) In general, some chromosomes in each generation have the highest degree of fitness. These chromosomes have the best genes and, due to having the best fitness, can produce better chromosomes for subsequent generations.
(2) This method increases the likelihood of the best parents moving to the next generation.

4.6. Fitness Function

The fitness function computes the effectiveness of selected solutions or created schedules. Our aim is to reduce the maximum flow time (F_{max}), so a low fitness is assigned to solutions which have the highest F_{max}. Here, the fitness function of the ith chromosome is $f(i) = 1\ \ F_{\max}(i)$ and $F_{max}(i)$ is the maximum flow time in chromosome i^{th}.

An example of editing the response sequence is shown in Figure 4 for calculating fitness.

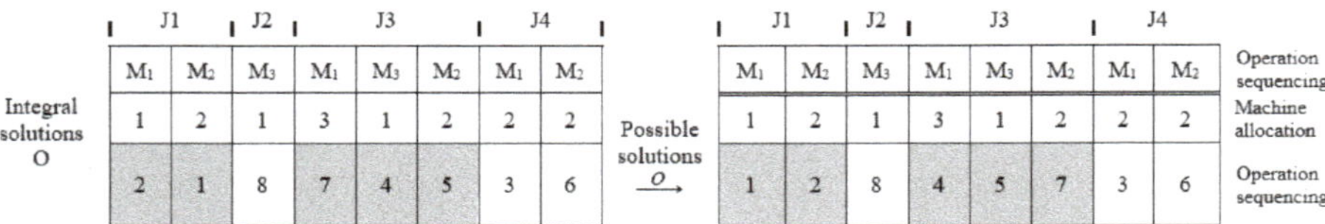

Figure 4. How to edit the sequence of operations.

4.7. Strategy for Dealing with Restrictions

The other issue in the GA is to deal with problem's restrictions, because genetic operators used in the algorithm may generate unjustified chromosomes. There is a corrective strategy in our algorithm to deal with these restrictions. In this strategy, an unjustified chromosome is transformed to a justified chromosome instead of being removed (Figure 3).

4.8. Termination of Genetic Algorithm

The algorithm of Figure 5 will terminate after max-gen repetitions.

```
Procedure of proposed GA
1.  Start
2.      Population counter's index (t) is 0
3.      Design the first dynamic two detention chromosome
4.      Repeat until termination condition is satisfied
5.        Create the new child C(t) by applying the crossover on P(t)
6.        Compute the fitness of child C(t)
7.        The C (t) parent P (t + 1) to determine now;
8.        Increase t for one point and test the condition on line 4
9.      End
10. End
```

Figure 5. Procedure of the proposed genetic algorithm.

5. Numerical Experiments

5.1. Comparison Method

Since the FDJSPM problem has not yet been considered, there is no similar algorithm to compare the proposed method with. So, in order to show the superiority of the proposed algorithm, we use the Random Keys Genetic Algorithm (RKGA) algorithm which was proposed in 2004 for the FSPM problem [37,38]. RKGA is a meta-heuristic algorithm which is designed for the FSPM problem with flexibility due to parallel machines with uniform speeds and the flexibility of operations in workstations. For this reason, the problem solved by this algorithm has the highest similarity to the FDJSPM problem. We can use it easily and without losing the generality. So, we consider the JSPM problem as a special case of the FDJSPM problem.

The developers of the RKGA algorithm used a hierarchical approach to solve their problem and decompose their problem into two sub-problems; allocation (task allocation to flexible workstations) and sequencing (sequence of processing). they then decomposed the allocation sub-problem into two sub-problems: workstation allocation and machine allocation. The allocation sub-problem is generated because of the flexibility of operations and parallel machines.

In studies conducted in 2004, for task scheduling, the RKGA method was used in the first step and then SPT Cyclic Heuristic (SPTCH) contributions and the Johnson rule were used for assigning tasks to machines in such a way that each task was allocated to a machine which can process the allocated task at the earliest time. In fact, in this algorithm, the first tasks are scheduled by RKGA, and then contributions are used to schedule the following operations [38]. We also use this equivalent concept for the FDJSPM problem. It means that the first operations of tasks are scheduled by RKGA and the following contributions are used to schedule the later operations [39].

5.2. Generate Random Problems

To generate random problems, 6 parameters are identified as Table 1. For the first five parameters, the RKGA algorithm is used [37,38] and U (1, 3) is considered as the processing speed of machines.

In general, all combinations of this level will be tested. Some other restrictions are introduced below. For machine distribution agents, it is necessary that at least one machine stage is different from the others. Also, the maximum number of machines in a stage must be lower than the number of tasks. At least one stage must have a number of parallel machines larger than 1. So, with 10 machines at each stage and 6 jobs removed, 6 compound machines and 6 jobs can be added at any stage. Ten datasets for each test scenario were generated.

Table 1. Level of agents to perform the GA.

Factors	Factors Levels			
Number of tasks	100-30		6	
Distribution machines	Constant	Variable	Constant	Variable
Number of machines	10-2	U(1, 4)–U(1, 10)	6-2	U(1, 4)–U(1, 6)
Number of steps	8-4-2			
Processing time	U(50, 70)–U (20, 100)			
Speed processing machines	U (1, 3)			

5.3. Design of Experiments

Algorithms (both proposed algorithm and RKGA algorithm) are coded in C++ language and were run on a computer with a Pentium IV CPU, 3 GHz, and 1GB of RAM and in Borland C^{++} 5.02. Each algorithm (proposed GA and RKGA) runs with 720 sets of information.

5.4. Setting Parameters

Different parameter values have a significant impact on the quality of the obtained solutions by the GA.

Here we only use the parameter setting procedure to set three parameters (p_m), (p_c), (μ'). For this purpose, one set is selected from each scenario (there are 10 data sets in each scenario). In total, 72 datasets are selected randomly, and then in the parameter setting procedure, the GA will be run on 72 datasets by creating new parameters. The average fitness is calculated for each run. First, we consider the following parameter values:

The percent of selection of the best chromosome for the next generation (μ'): 5%, 10%, 15%, 20%, 25%.

Probability of crossover (p_c): 0.3, 0.4, 0.5, 0.6, 0.7, 0.8, 0.9

Probability of mutation (p_m): 0.005, 0.01, 0.015, 0.02, 0.025

Number of initial population (pop-size): 100

Number of generations (Max-gen): 125

In the parameter setting procedure, at first, we change the value of one parameter in its defined domain, then fix the other parameters in their minimum value, and finally define the best for the changeable parameter based on their fitness, and then we change the other parameter in the next stage by fixing this value for the mentioned parameter. Of course, in addition to these fixed parameters, other parameters are kept at their minimum values and, as for the previous stage, the best value is selected for changeable parameters. This procedure is repeated until all parameters are fixed. Computational results show the best-obtained values for (μ') as 20%, 0.7 and $(p_m)w$ (p_c), 0.02 respectively. (Table 2).

Table 2. Values of GA parameters after setting.

p_m	p_c	μ'	Number of Generations	Size of Initial Population	Parameter
0.02	0.7	20%	125	100	Value

6. Computational Results

In this section, the performance of the proposed GA algorithm for the FDJSPM problem with the aim of reducing the maximum flow time will be compared to RKGA. The comparison was performed for 216 scenarios and was run 10 times as shown in Table 3.

Table 3. Comparison of the proposed GA and the RKGA.

Magnitude of the Problem		Small				Medium				Large			
Number of Jobs		6				30				100			
Number of Steps		2	4	8	Average	2	4	8	Average	2	4	8	Average
RKGA [30,31]	Lowest F_{max}	40/36	73/93	172/6	95/62	91/85	122/6	204	139/47	283/9	307/1	362/7	317/89
	Average F_{max}	121/09	184/65	302/83	202/86	527/43	646/83	845/23	673/16	1743/80	1905/88	2612/90	2087/53
Proposed GA	Lowest F_{max}	39/74	74/95	169/86	94/85	93/35	118/01	202/02	137/79	278/12	307/98	365/29	314/13
	Average F_{max}	119/66	184/71	295/67	200/01	527/69	635/21	825/87	662/92	1731/32	1870/89	2562/59	2054/93
Improvement (%)	Lowest F_{max}	1%/54	−1/38%	1%/56	0%/57	%−1/63	3%/73	0%/96	1%/02	2%/02	%−0/28	1%/76	1%/17
	Average F_{max}	1%/17	−0/03%	2%/37	1%/17	%−0/05	1%/80	2/29	1%/35	0%/72	1%/84	1%/93	1%/49
Improvement Count	* Lowest F_{max}	58	29	63	50/0	23	66	51	46/7	62	41	53	52/0
	** Average F_{max}	12	7	18	13/3	10	19	13	14/0	18	15	16	16/3
Solving Time Improvement (s)		−0/4	−0/75	−3/1714	−1/4	−7/6567	−7/6774	−33/258	−14/5	−0/65605	−11/0858	−89/6802	−33/8

*: The lowest F_{max} in 70 runs of scenarios based on the number of stages. The values are not identical with RKGA.
: The lowest value of average on F_{max} in 21 runs of scenarios based on the number of stages. Their values are not identical with RKGA. **Note: Negative numbers show the superiority of RKGA method against the proposed GA algorithm.

A main index named "the average minimization of the maximum flow time of tasks" is used to compare these methods and other auxiliary indexes, such as "the minimum value of minimization of the maximum flow time of tasks during different runs", "average time to solve each scenario", and "the fold improvement in objective function" and different size of problem (small, medium, and large) are used. The results of this comparison, according to Tables 4 and 5, are as follow:

1. According to the "the minimum value of the minimization of the maximum flow time of tasks" index, and for 10 runs of each of the 72 scenarios of different sizes (small, medium, large), the proposed GA has an average improvement of 0.57%, 1.02%, and 1.17%, and an overall improvement of 0.92% above RKGA.
2. "Average objective function" for 10 runs on 72 datasets seems the best criterion for evaluating these algorithms. As seen in Table 3, according to this criterion for small, medium, and large size problems, the proposed algorithm has 1.17%, 1.35%, and 1.49% average improvement and an overall improvement of 1.34% higher than RKGA.
3. According to the "improvement of objective function" index for small, medium, and large size problems, the proposed algorithm has 50, 46.7, and 52-fold improvement, respectively, and on average has a 49.6-foldimprovement over RKGA.
4. According to "average objective function" index for small, medium, and large size problems, the proposed algorithm has an average improvement of 40, 42, and 49-fold from 72 runs, respectively, and, overall, has an average improvement of 49.6% above RKGA.
5. According to "average time for solving the problem", the proposed algorithm has no significant improvement for small problems, but for medium and large problems, it has an average increase of 14.5% and 33.8%, respectively, and has worse performance than RKGA.

"Average difference between GA and RKGA" is a suitable criterion to show the performance of the proposed algorithm. For this index, there are 1.17%, 1.35%, and 1.49% improvements (for small, medium, and large problems) for the proposed algorithm. This result shows with increasing the size of the problem, the proposed algorithm has greater improvement than RKGA.

Finally, the results show that the proposed GA requires more time to solve the problem, but with respect to the improvements obtained from minimum and average on the objective function, this increased time is justified. Therefore, according to the numerical results, the effectiveness of the proposed algorithm against the RKGA is confirmed.

Figure 6 shows the Average F_{max} comparison for the two algorithms.

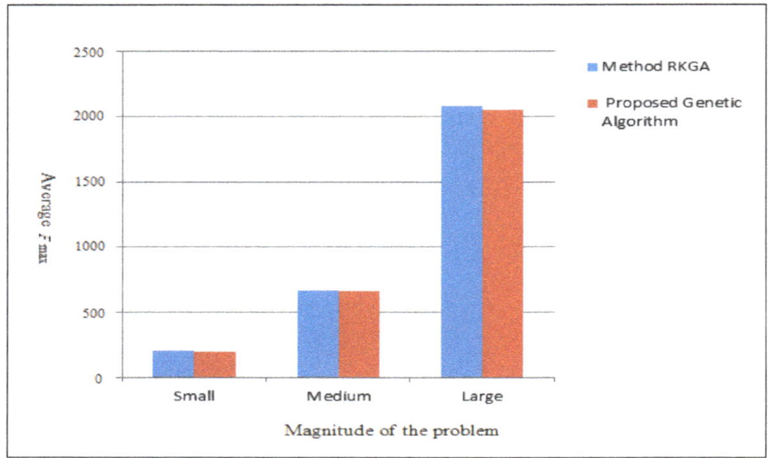

Figure 6. Average F_{max} Comparison.

Table 4. Average solving time of the proposed GA and RKGA.

RKGA	Small Dimension	Medium Dimension	Large Dimension	Sum
The total number of smaller improvements of F_{max}	150	140	156	446
Total percent of the improvement of F_{max}	20%/83	19%/44	21%/67	20%/65
GA (proposed)	**Small Size**	**Medium Size**	**Large Size**	**Addition**
Total numbers of improvement on F_{max}	40	42	49	131
Total percent of the improvement on F_{max}	55%/56	58%/33	68%/06	60%/65

Table 5. Improvement and percent of improvement between the proposed GA and RKGA.

Size of Problems	Small				Medium				Large			
Number of Tasks	6				30				100			
Number of Stages	2	4	8	Average	2	4	8	Average	2	4	8	Average
RKGA method	025/0	028/0	035/0	029/0	201/0	248/0	233/0	227/0	602/1	551/1	745/1	633/1
Proposed genetic algorithm	035/0	049/0	146/0	077/0	735/0	152/2	982/7	623/3	653/2	745/18	237/158	88/59

Investigating the results obtained from the proposed algorithm and comparing it with the RKGA method shows that the proposed method is slightly better than the previous one. For big problems, this gap is slightly higher. The main reason for this result is the use of intelligent intersection operators and mutants that can search the response space more effectively than RKGA.

7. Conclusions

In this paper, the flexible job-shop scheduling problem with parallel machines (with non-uniform speed) in each workstation for dynamic manufacturing systems (because time entrance of tasks is variable) is introduced. Then, a mixed integer nonlinear programming model is presented. Then, F_{max}

as an objective function is investigated, with the aim of the efficient utilization of machines by reducing the time that recourses are used for, thus increasing the rate and speed of the manufacturing process, which is the most important goal in the manufacturing supply chain. We utilize the capability of a GA to solve the time and scheduling problems. Due to the dynamicity of manufacturing, dynamic two-dimensional chromosomes are provided. Chromosomes and genetic operators are designed in such a way that there is minimal dealing with constraints and unjustified chromosomes. In the proposed GA, the initial population is generated randomly and doesn't have any information about the problem. This maintains the diversity of the population and reduces the possibility of premature convergence. Also, the influence of strong chromosomes is reduced and premature convergence to the local optimum is avoided by preventing the entrance of duplicate chromosomes into the next generation. After setting our proposed genetic parameters, its performance was compared to the RKGA method and it was observed that our genetic algorithm requires more time to solve the problem than the RKGA method. However, this time allows improvements in the minimum and average value of the objective function. Computational results show improvements of 1.17%, 1.35% and 1.49, and 1.34% on average, for small, medium, and large problems. Thus, according to the numerical experiments, the efficiency of the proposed algorithm is confirmed as greater than for RKGA. Increasing the efficiency of proposed algorithm and reducing the execution time are considered as future goals. On the other hand, in actual manufacturing issues, decision makers are often faced with multiple objectives. Selecting the right combination of objectives and investigating the problem in a multi-objective case and in a stochastic environment is essential. As future work, our algorithm could be integrated with other methods. For example, using a local search algorithm can improve the performance of the GA.

Author Contributions: A.K.S. has conceived, designed and formulated the complete paper; M.Y.S. and M.S. helped to design the algorithms; S.M.B. and A.A.R.H. have investigated data analytics and formulated the methodology of a paper; J.W. improved the algorithms, performance analysis and results.

Acknowledgments: This work is supported by the National Natural Science Foundation of China (61772454, 61811530332, 61811540410, U1836208).

Conflicts of Interest: The authors declare no conflict of interest.

References

1. Baker, K.R. *Introduction to Sequencing and Scheduling*; John Wiley & Sons: New York, NY, USA, 1974.
2. Hosseinabadi, A.R.; Siar, H.; Shamshirband, S.; Shojafar, M.; Nasir, M.H.N.M. Using the gravitational emulation local search algorithm to solve the multi-objective flexible dynamic job shop scheduling problem in Small and Medium Enterprises. *Ann. Oper. Res.* **2015**, *229*, 451–474. [CrossRef]
3. Tay, J.C.; Ho, N.B. Evolving dispatching rules using genetic programming for solving multi-objective flexible job-shop problems. *Comput. Ind. Eng.* **2007**, *54*, 453–473. [CrossRef]
4. Shamshirband, S.; Shojafar, M.; Hosseinabadi, A.R.; Kardgar, M.; Nasir, M.H.; Ahmad, R. OSGA: Genetic-based open-shop scheduling with consideration of machine maintenance in small and medium enterprises. *Ann. Oper. Res.* **2015**, *229*, 743–758. [CrossRef]
5. Farahabadi, A.B.; Hosseinabadi, A.R. Present a New Hybrid Algorithm Scheduling Flexible Manufacturing System Consideration Cost Maintenance. *Int. J. Sci. Eng. Res.* **2013**, *4*, 1870–1875.
6. Hosseinabadi, A.R.; Farahabadi, A.B.; Rostami, M.S.; Lateran, A.F. Presentation of a New and Beneficial Method Through Problem Solving Timing of Open Shop by Random Algorithm Gravitational Emulation Local Search. *Int. J. Comput. Sci.* **2013**, *10*, 745–752.
7. Shojafar, M.; Kardgar, M.; Hosseinabadi, A.R.; Shamshirband, S.; Abraham, A. TETS: A Genetic-based Scheduler in Cloud Computing to Decrease Energy and Makespan. In Proceedings of the 15th International Conference on Hybrid Intelligent Systems (HIS 2015), Seoul, Korea, 16–18 November 2015; Chapter Advances in Intelligent Systems and Computing. Volume 420, pp. 103–115.
8. Gena, M.; Zhang, W.; Line, L.; Yun, Y.S. Recent advances in hybrid evolutionary algorithms for multiobjective manufacturing scheduling. *Comput. Ind. Eng.* **2017**, *112*, 616–633. [CrossRef]

9. Wang, J.; Ju, C.; Gao, Y.; Sangaiah, A.K.; Kim, G. A PSO based Energy Efficient Coverage Control Algorithm for Wireless Sensor Networks. *Comput. Mater. Contin.* **2018**, *56*, 433–446.
10. Tirkolaee, E.B.; Hosseinabadi, A.R.; Soltani, M.; Sangaiah, A.K.; Wang, J. A Hybrid Genetic Algorithm for Multi-trip Green Capacitated Arc Routing Problem in the Scope of Urban Services. *Sustainability* **2018**, *10*, 1366. [CrossRef]
11. Chen, J.C.; Wub, C.C.; Chen, C.W.; Chen, K.H. Flexible job shop scheduling with parallel machines using Genetic Algorithm and Grouping Genetic Algorithm. *Expert Syst. Appl.* **2012**, *39*, 10016–10021. [CrossRef]
12. Dalfard, V.M.; Mohammadib, G. Two meta-heuristic algorithms for solving multi-objective flexible job-shop scheduling with parallel machine and maintenance constraints. *Comput. Math. Appl.* **2012**, *64*, 2111–2117. [CrossRef]
13. Vokurka, R.J.; O'Leary-Kelly, S.W. A review of empirical research on manufacturing exibility. *J. Oper. Manag.* **2000**, *18*, 85–501. [CrossRef]
14. Naderi, B.; Azab, A. Modeling and heuristics for scheduling of distributed job shops. *Expert Syst. Appl.* **2014**, *41*, 7754–7763. [CrossRef]
15. Shahrabi, J.; Adibi, M.A.; Mahootchi, M. A reinforcement learning approach to parameter estimation in dynamic job shop scheduling. *Comput. Ind. Eng.* **2017**, *110*, 75–82. [CrossRef]
16. Yazdani, M.; Aleti, A.; Khalili, S.M.; Jolai, F. Optimizing the sum of maximum earliness and tardiness of the job shop scheduling problem. *Comput. Ind. Eng.* **2017**, *107*, 12–24. [CrossRef]
17. Riane, F.; Artiba, A.; Elmaghraby, S.E. A hybrid three-stage flowshop problem: Efficient heuristics to minimize makespan. *Eur. J. Oper. Res.* **1998**, *109*, 321–329. [CrossRef]
18. Nowicki, E.; Smutniciki, C. The flow shop with parallel machines: A taboo search approach. *Eur. J. Oper. Res.* **1998**, *106*, 226–253. [CrossRef]
19. Brandimarte, P. Theory and methodology, exploiting process plan flexibility in production scheduling: A multi-objective approach. *Eur. J. Oper. Res.* **1999**, *114*, 59–71. [CrossRef]
20. Ghedjati, F. Genetic algorithms for the job-shop scheduling problem with unrelated parallel constraints: Heuristic mixing method machines and precedence. *Comput. Ind. Eng.* **1999**, *37*, 39–42. [CrossRef]
21. Jansen, K. Approximation algorithms for flexible job shop problems. *Int. J. Found. Comput. Sci.* **2001**, *12*, 521–534.
22. Kacem, I.; Hammadi, S.; Borne, P. Approach by localization and multi objective evolutionary optimization for flexible job-shop scheduling problems. *IEEE Trans. Syst. Man Cybern.* **2002**, *32*, 1–13. [CrossRef]
23. Lee, Y.H.; Jeong, C.S.; Moon, C. Advanced planning and scheduling with outsourcing in manufacturing supply chain. *Comput. Ind. Eng.* **2002**, *43*, 351–374. [CrossRef]
24. Kim, Y.K.; Park, K.; Ko, J. A symbiotic evolutionary algorithm for the integration of process planning and job shop scheduling. *Comput. Oper. Res.* **2004**, *30*, 1151–1171. [CrossRef]
25. Allet, S. Handling flexibility in a "generalised job shop" with a fuzzy approach. *Eur. J. Oper. Res.* **2003**, *147*, 312–333. [CrossRef]
26. Scrich, C.A.; Armentano, V.A.; Laguna, M. Tardiness minimization in a flexible job shop: A taboo search approach. *J. Intell. Manuf.* **2004**, *15*, 103–115. [CrossRef]
27. Gen, M.; Cheng, R. *Genetic Algorithm in Search, Optimization and Machine Learning*; Addition-Wesley: Reading, MA, USA, 2004; ISBN 0201157675.
28. Low, C. Simulated annealing heuristic for flow shop scheduling problem with unrelated parallel machines. *Comput. Oper. Res.* **2005**, *32*, 2013–2025. [CrossRef]
29. Xia, W.; Wu, Z. An effective hybrid optimization approach for multi-objective flexible job-shop scheduling problems. *Comput. Ind. Eng.* **2005**, *48*, 409–425. [CrossRef]
30. Torabi, S.A.; Karimi, B.; Fatemi-Ghomi, S.M.T. The common cycle economic lot scheduling in flexible job shops: The finite horizon case. *Int. J. Prod. Econ.* **2005**, *97*, 52–65. [CrossRef]
31. Wang, J.X.X. Complexity and algorithms for two-stage flexible flow shop scheduling with availability constraints. *Comput. Math. Appl.* **2005**, *50*, 1629–1638.
32. Tay, J.C.; Wibowo, D. An effective chromosome representation for evolving flexible job-shop scheduling. In Proceedings of the Genetic and Evolutionary Computation Conference (GECCO), Seattle, WA, USA, 26–30 June 2004; pp. 210–221.
33. Kyparisis, G.J.; Koulamas, C. Flexible flow shop scheduling with uniform parallel machines. *Eur. J. Oper. Res.* **2004**, *168*, 985–997. [CrossRef]

34. Ho, N.B.; Tay, J.C.; Lai, E. An effective architecture for learning and evolving flexible job shap schedules. *Eur. J. Oper. Res.* **2007**, *179*, 316–333. [CrossRef]
35. Gao, J.; Gen, M.; Sun, L. Scheduling jobs and maintenances in flexible job shop with a hybrid genetic algorithm. *J. Intell. Manuf.* **2007**, *17*, 493–507. [CrossRef]
36. Park, B.J.; Choi, H.R.; Kim, H.S. A hybrid genetic algorithm for the job shop scheduling problems. *Comput. Ind. Eng.* **2003**, *45*, 597–613. [CrossRef]
37. Kurz, M.E.; Askin, R.G. Scheduling flexible flowlines with sequence-dependent setup times. *Eur. J. Oper. Res.* **2004**, *159*, 66–82. [CrossRef]
38. Kurz, M.E.; Askin, R.G. Comparing scheduling rules for flexible flow-lines. *Int. J. Prod. Econ.* **2003**, *85*, 371–388. [CrossRef]
39. Abbasian, M.; Nahavandi, N. Minimization Flow Time in a flexible Dynamic Job Shop with Parallel Machines. Master's Thesis, Engineering Department of Industrial Engineering, Tarbiat Modares University, Tehran, Iran, 2008.

Article

Design of a Symmetry Protocol for the Efficient Operation of IP Cameras in the IoT Environment

Jaeseung Lee [1], Jungho Kang [2], Moon-seog Jun [1] and Jaekyung Han [3,*]

1 Department of Computer Science and Engineering, Soongsil University, Seoul 07027, Korea; ljs0322@ssu.ac.kr (J.L.); mjun@ssu.ac.kr (M.-s.J.)
2 Department of Information Security, Baewha Women's University, Seoul 03039, Korea; kjh7548@naver.com
3 Department of Construction Legal Affairs, The Graduate School of Construction Legal Affairs, Kwangwoon University, Seoul 01890, Korea
* Correspondence: hjk1014@kw.ac.kr; Tel.: +82-29-405-788

Received: 8 January 2019; Accepted: 28 February 2019; Published: 11 March 2019

Abstract: The rapid development of Internet technology and the spread of various smart devices have enabled the creation of a convenient environment used by people all around the world. It has become increasingly popular, with the technology known as the Internet of Things (IoT). However, both the development and proliferation of IoT technology have caused various problems such as personal information leakage and privacy violations due to attacks by hackers. Furthermore, countless devices are connected to the network in the sense that all things are connected to the Internet, and network attacks that have thus far been exploited in the existing PC environment are now also occurring frequently in the IoT environment. In fact, there have been many security incidents such as DDoS attacks involving the hacking of IP cameras, which are typical IoT devices, leakages of personal information and the monitoring of numerous persons without their consent. While attacks in the existing Internet environment were PC-based, we have confirmed that various smart devices used in the IoT environment—such as IP cameras and tablets—can be utilized and exploited for attacks on the network. Even though it is necessary to apply security solutions to IoT devices in order to prevent potential problems in the IoT environment, it is difficult to install and execute security solutions due to the inherent features of small devices with limited memory space and computational power in this aforementioned IoT environment, and it is also difficult to protect certificates and encryption keys due to easy physical access. Accordingly, this paper examines potential security threats in the IoT environment and proposes a security design and the development of an intelligent security framework designed to prevent them. The results of the performance evaluation of this study confirm that the proposed protocol is able to cope with various security threats in the network. Furthermore, from the perspective of energy efficiency, it was also possible to confirm that the proposed protocol is superior to other cryptographic protocols. Thus, it is expected to be effective if applied to the IoT environment.

Keywords: IP camera; IP camera security; NVR security; video security

1. Introduction

As Internet technology has developed rapidly, and its penetration rate has increased greatly in recent years, the Internet environment now has a closer relationship with humans, and technology related with the so-called Internet of Things (IoT) is becoming increasingly widespread. However, as IoT technology develops and demand for it increases, various types of security incidents are taking place due to exploitation of the growing number of security vulnerabilities.

As the meaning of the IoT implies, the more devices that are connected to the network, the greater the number of network-based security attacks. Furthermore, the application of existing security solutions such as vaccines and firewalls—hitherto applied to PCs, servers and networks—to the IoT environment, has raised awareness of several limitations. Consequently, there is a growing need for device security that fits the IoT environment. As an example of a major security incident, numerous IoT devices were infected and exploited by a large-scale DDoS attack on the Internet domain service provider Dyn, which for several hours made it impossible to access dozens of popular websites such as The New York Times, Twitter, Netflix, and Amazon. The DDoS attack is not a new type of attack, but it is one that has been used most frequently to exploit and abuse the existing PC environment by attacking and infecting PCs beforehand and then exploiting them. By examining cases of such attacks, it was possible to confirm that smart devices used in the IoT environment, including DVRs, tablets, CCTVs (Closed Circuit Televisions) and other everyday smart devices, can be used for network attacks [1,2]. In addition, IoT devices connected to Internet Explorer can be hacked and exploited for attacks. A recent example of this attack was identified by Imperva, a company dedicated to security [3]. The attack was found to be a traditional HTTP flaw attack used in attacks on the Internet environment of existing PCs, which aimed to overload resources on the cloud service. However, what is particularly noteworthy about this attack is that it came from an IP camera rather than a traditional computer botnet. Imperva explained that this attack made up to 20,000 requests per second by exploiting about 900 CCTV cameras, using the embedded version of Linux and the BusyBox toolkit [4,5]. The attackers exploited the software vulnerabilities and the social engineering methods used to infect computers for attacks in the existing environment, but this attack was performed in an environment that could be attacked easily by accessing the Internet through Telnet and SSL [6]. Also, to attack the IoT devices, "root" and "admin", and "admin" and "password" were mainly used as the ID and password combinations, respectively. These combinations could be used to exploit the fact that the product default values had not been changed. In order to prevent the problems that can occur in the IoT environment, it is necessary to apply a security solution to the devices, but the situation is not easy to handle. In the IoT environment, it is difficult to install and execute security solutions due to the features of small devices with limited memory space and computational power, and it is also difficult to protect certificates and encryption keys due to easy physical access [7]. Therefore, in this paper, we discuss the trends and security problems of IP cameras used in the above cases in the IoT environment, and propose a secure method of communication by configuring the NVR network security design system.

2. Related Works

2.1. Internet of Things

The "Internet of Things" refers to intelligent technologies and services that can connect innumerable devices scattered widely across different areas to a single network to facilitate the exchange of information through communication. The technological concept that enables autonomous communication between humans and objects, as well as between objects, through human intervention, and provides services by analyzing the surrounding environment, is considered to be the leading technology of the digital revolution.

Since Kevin Ashton, the founder of the MIT Auto-ID Center, proposed both the concept of the IOT and the term itself in 1999, it has been undergoing continuous development and it now easy to encounter the IoT as many devices and services which have been developed accordingly.

Currently there is a wide range of IT products that use the IoT in the market. The IP camera field, which is the environment proposed in this paper, and many IoT services and devices are under development with a view to applying IOT technology in the fields of crime and disaster prevention. Objects must be able to communicate with each other in such an environment, and machine-to-machine communication is the base technology that supports such interaction.

However, most IoT services and devices currently operate only within the domain of the same manufacturer or service. Therefore, a standardized mode is essential to establish machine-to-machine communication, especially secured communication, between objects that are made by different manufacturers and used in different business domains. Moreover, the existing method of communication requires SSL and RSA security protocols, which were developed for the existing network environment, and these may not be efficient in an environment such as the IP camera network, which has certain battery and storage limitations. Therefore, this paper proposes a protocol that is both stable and more energy efficient than existing security protocols, and hence is applicable to the IP camera environment.

2.2. IP Camera Trends

As the term Closed-Circuit Television implies, existing CCTV systems have their own closed configurations for the purpose of security. Therefore, it is common to install CCTV cameras in a specific area to monitor captured images at predetermined locations, and to store and retrieve them when necessary. On the other hand, the development of information technology (IT) has led to the emergence of IP cameras, which have the advantage of being open, scalable and flexible, unlike conventional CCTVs, by utilizing the concept of transmitting images based on the networks. Because they are linked to the Internet via a network connection, IP cameras have a great feature in that they can be used anytime and anywhere, provided that the Internet is available [8]. This feature is advantageous in that the installation and configuration of a network is both simple and inexpensive in the present era, and in many cases networks have been installed and configured in most places, so that it is relatively easy to use IP cameras for everyday purposes at no additional cost. In addition, while conventional analog CCTV cameras capture video only, the latest IP cameras are equipped with functions for compressing and transmitting video and audio at the same time, and most IP cameras also support voice recording and transmission. This not only makes it possible to expand from conventional simple video surveillance to the communication function, but also makes it possible to utilize them for security surveillance using voice [9].

If one looks at the types of IP cameras currently being released, they are very different from conventional CCTV cameras. First, they typically come in special shapes such as the box, dome, bullet, and flat rectangular types. Furthermore, they can be classified into fixed pan & tilt, zoom and high-speed dome types according to camera module movement, and into vandal-proof, weather-proof and waterproof types according to their proofing function. In addition, they usually feature an infrared (IR) light emitting diode (LED), which is either fitted on the front of the camera or attached separately for operation in nighttime and low-light environments. In addition, WDR (Wide Dynamic Range), is also becoming more common. In the early stage of transition from CCTVs to IP cameras, the terminology was not yet established, so the term IP camera was used along with web camera and network camera. In this paper, the target device is also referred to as a "network camera", but in recent years this has almost been unified with the term "IP camera", while "web camera" specifically indicates a product composed only of a camera module connected to a PC via a USB [10].

2.3. IP Camera Market Trends

The global video surveillance market was worth $17.2 billion in 2017, and is expected to grow at an average annual rate of 7.2% to reach $22.7 billion by 2021. Both economic growth and demand for security in emerging countries are expected to continue rising as a key driving force in the growth of the global video surveillance market. The most direct cause is related to attacks by global terrorism and the resulting increase in civilian deaths. Since 2012, the number of people killed by terrorism has steadily increased in the international community. In 2015, the number of terrorist attacks worldwide came to 11,770 and the death toll reached 28,320, representing increases of 14.5% and 26.0% respectively compared to 2012. To prevent these problems, demand for video surveillance systems by governments in each country is increasing, and governments and intelligence agencies are increasingly deploying such systems to curb criminal activity and control hostile situations.

This trend is also associated with the development of IoT technology, which is making system installation and management much easier. IoT-based connection devices are being applied to video surveillance systems, and the number of connected devices in the world is expected to increase sharply from 8.7 billion in 2012 to 22.9 billion in 2016. As the numbers of IoT-connected devices and IoT access control solutions are increasing explosively, the related video surveillance market is expanding accordingly [11].

In addition, the world's urban population increased from 3.94 billion in 2012 to 4.02 billion in 2016, and this increase has led to a greater need for advanced safety products and services to ensure a safer urban life, prevent crime and promote disaster and incident preparedness. Also, as the urban population increases, the need for high-tech safety products is being widely emphasized. As a result, the demand for associated services is increasing and improvements in the quality and performance of safety products and services are rapidly progressing [12].

We can explain the cause of this increase in terms of the social environment as well as technology. The number of Internet users worldwide is expected to increase to 3.48 billion in 2016, up 42.0% from the 2.45 billion recorded in 2012. The increase in the number of users of wired and wireless Internet is also expected to facilitate the increasing demand for IP cameras. The improved accessibility of computers and mobile devices to wired and wireless networks has increased the Internet usage rate and the number of Internet users, and this increase is a key factor in the rising demand for IP cameras.

Worldwide per capita GDP in 2016 was $10,337, up 5.7% from $9781 in 2012, and economic growth in both developed and developing countries is driving investment in video surveillance products and services. The growth of global GDP has resulted in an increase of consumer purchasing power for video surveillance products and services, and an increase in demand for security and safety related products and services, and is leading to the growth of worldwide video surveillance markets [13].

2.4. Cases of IP Camera Security

The growth of the IP camera market has not only had a positive impact in such areas as anti-terrorism, crime prevention, disaster and incident monitoring, but also a negative impact. Over the past several years, malicious attacks on IP cameras have caused plenty of problems. Austrian security researchers have identified more than 80 backdoors to Sony's IP cameras, while Israeli security experts have found vulnerabilities in IP camera models. SEC Consult, an Austrian security company, has found two different kinds of backdoors in Sony's IPELA Engine IP cameras. According to this company, it is possible to remotely control "primana" and "debug" via Telnet, so their users are recommended to update the firmware through the SNC Toolbox to fix the vulnerabilities.

In addition, Cybereason purchased twelve types of IP cameras from eBay and Amazon, and found that all the passwords for the purchased devices were "888888", which made it difficult to defend their firewalls against attackers. It was a vulnerable situation, in which the companies that made such IP cameras had not updated the firmware.

Security vulnerability has also been detected in SWANN's cameras, which showed that they can intercept messages sent from OzVision's computer server to the Swann camera app with a free tool commonly used in the security industry. The Swann Camera app is used to view images taken by the camera on a smartphone. The intercepted message includes a serial number for each camera given by the factory. By changing the serial number, a research group composed of five European security consultants was able to acquire the images from other cameras. The researchers succeeded in acquiring the images simply by inserting the serial number of the camera they had purchased. During this process, they did not have to enter the ID or password of another user account. They also found a way to identify the serial number used by Swann's camera. Figure 1 is a program action screen that changes the actual serial number. Theoretically, it is possible to steal images quickly from any account, causing a major problem with personal image information leakage.

There was also a case in which a DDoS attack used IP cameras. On 21 October 2016, the DNS service provider Dyn was subjected to a massive Distributed Denial of Service (DDoS) attack, which resulted in the simultaneous paralysis or delay of seventy-six sites including Twitter, Netflix, and The New York Times (NYT).

According to the Bryan Krebs blog (Krebs on Security) analysis, it was confirmed that the cause of the DDoS attack was a vulnerable IoT device (i.e., a device whose factory default ID/PW was not reset), infected with the Mirai malicious code, which executed the DDoS attack initiated by hackers. Most IoT devices are connected online with a weak factory default ID/PW configuration, making them vulnerable to attacks. Therefore, the number of attempts to infect malicious code targeting such devices has been increasing every year, making security measures essential. Moreover, the Mirai malicious codes used in the abovementioned example of large-scale DDoS attacks have been made public, thus enabling anyone to make a malicious code easily, by simply changing the source codes. As a result, we are faced with the possibility not only of large-scale DDoS attacks, but also an Internet crisis in which IoT devices are exploited by new malicious codes capable of using the source codes [14,15].

The manufacturers of IoT equipment (IP cameras, Internet routers, set-top boxes, etc.) use various CPUs (ARM, MIPS, PowerPC, SuperH, etc.) and adopt the Linux operating system, which is suitable for such a CPU environment. Based on the Linux operating system, the source codes are made to be executable in various CPU environments through cross-compilation. On this account, almost all IoT equipment is subject to attacks. We can confirm that the malicious codes actually found had the same functionality, but were designed to run in various CPU environments such as ARM, MIPS, and PowerPC [16].

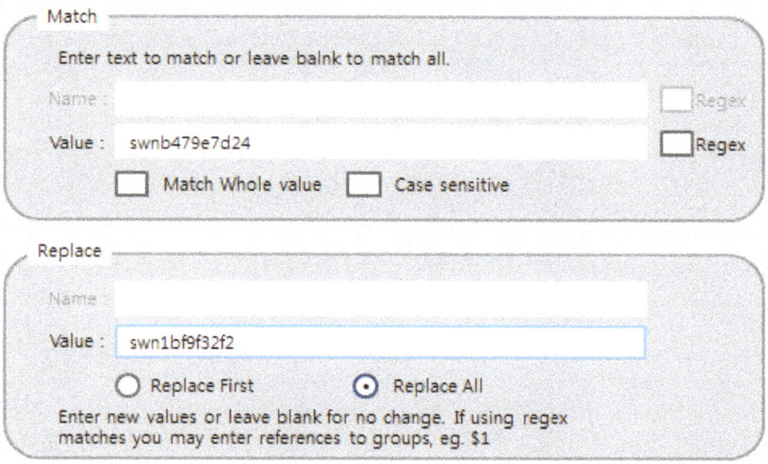

Figure 1. Picture of the serial number change program.

2.5. Network Attack Methods

2.5.1. DDoS

The Distributed Denial of Service attack is a method of performing multiple DOS attacks in parallel. It consists of creating a Denial-of-Service situation to prevent users from using a given service by generating heavy traffic that cannot be accommodated in the operating servers and network equipment of the company providing the service. The main characteristic of the DDoS attack is that the principal objective is to stop a system service for a certain period of time in order to prevent it from being provided to users, rather than hacking, information leakage or taking control of the server systems to acquire the highest privileged account [17,18].

The target servers attacked by an attacker are not damaged by the deletion, modification, leakage or destruction of data, but there may arise file system or other damage due to the system attack. This implies that if used in combination with other types or methods of attack, it can be a pre-work of system paralysis for effective intrusion. It is difficult to trace the cause of a DDoS attack and the attacker. In the case of a DoS attack, there are many ways to falsify the attack sources, and for a distributed DoS attack, even if it is possible to track an attack host, it is still difficult to find out when the host was occupied or what unexpected route was used for the takeover, and to keep track of them, and so on.

2.5.2. Sniffing and Spoofing

Sniffer was originally a registered trademark of Network Associate, but is now used as a general term, like PC or Kleenex. It is a tapping device that eavesdrops on traffic flows within the computer network. In the case of a sniffing attack, it has become a very threatening type of attack for companies, such as web hosting service providers and IDC centers connected in the same network. If an attack on a single system succeeds, the attacker will be able to intercept the entire data flow on the network through the system and thereby acquire sensitive information, such as user IDs and passwords [19].

In the case of the Ethernet, all hosts in the local network share the same communication line, which means that any computer on the same network can observe every different computer's traffic. However, if one's computer allows all traffic passing through the Ethernet, it must deal with irrelevant and unnecessary traffic, which is inefficient and results in poor network performance. Therefore, an Ethernet interface should have a filtering function that ignores irrelevant traffic that does not include the user's own MAC address, thus processing only traffic that has the user's own MAC address. Nevertheless, the user can set a function to observe all traffic on the Ethernet interface. This is called the promiscuous mode. The sniffer will set the Ethernet interface to the promiscuous mode and thus be able to eavesdrop on all traffic going through the local network.

Spoofing is a method of hacking used by malicious attackers who build a fake website and induce users to visit it, and thereby obtain the system privileges of users who access the real website and steal their information by exploiting the structural flaws of the Internet Protocol (TCP/IP). ARP Spoofing is an attack scheme that exploits flaws in the ARP protocol to cheat its MAC address as the MAC addresses of other users. By exploiting the vulnerability of ARP Request Broadcasting, attackers can obtain accurate information about all Host IP-MAC address mappings on the network [20,21].

2.6. DVR and NVR

The DVR (Digital Video Recorder) is a device that converts image data input from an analog camera into high-quality digital images using a capture board and stores them on a hard disk. The DVR converts recorded images into digital images and stores them on the hard disk semi-permanently, allowing users to search them according to their recorded data condition (event, date, etc.). It is also a multifunctional digital recording and monitoring device equipped with an image transmission function that can search recorded images and monitor screens in real time using LAN, MODEM, ADSL, etc. at a remote place from a long distance [22,23].

The NVR (Network Video Recorder) is a system that receives and stores video data from the cameras, videos, and servers installed on a network. It has both networking and the ability to store received video data in real time, and to decode and output them to a monitor [24,25]. The number of connected cameras can be limited according to the resolution of the received image. The NVR performs recording and playback simultaneously. Video data recorded on a single device can be remotely viewed by several authorized operators scattered throughout the network. They are independent and do not affect each other. In addition, it is easy to expand the number of NVR units, and even if you have plenty of NVR units throughout the system, you only have to connect it to the network to add one more NVR. In this paper, we propose a security system that is based on the NVR system [26,27].

3. Contents of the Proposed System

This chapter presents a design method of a control center to cope with security threats and to operate IP cameras effectively. As they are connected to the network, IP cameras combined with the NVR system described above are vulnerable to various security threats such as user authentication and access security threats that can occur in the network. IP cameras are currently being used not only in numerous homes and offices, but also in disaster, incident and crime prevention. ID/PW authentication can be taken easily by indiscriminate insertion attacks. As such, this paper proposes a secure authentication protocol using the user code, random data, and the real-time request time value from the standpoint of IP camera users and the control center.

In this paper, the procedure is divided into the following three steps: First, the user registers an IP camera. In order to utilize an IP camera, its ID/PW-based registration procedure for user registration is performed in the control center. During this process, a transfer process is performed for the serial number and random value of the product as well as the polynomial expression to be used in the authentication process later on. The second step involves a procedure for viewing the images stored in the control center to be performed. After proceeding to the authentication of the user, the control center and the IP camera, viewing is allowed if the user is judged to be suitable. The third step consists of the real-time monitoring of IP cameras. The user is proved to be a legitimate user to connect an IP camera and receives video images in real time. The proposed protocol applies the group key method between the devices, the control center and the users, to enable secure communication. In addition, the proposed protocol is designed in such a way that the more IoT devices (including IP cameras) are connected to the network, the more reliable the authentication procedure becomes. Thus, a strong authentication system is constructed to reflect the current tendency to use multiple IoT devices. The proposed entire protocol procedure is as shown in Figure 2, Table 1 is the meaning of Notation where Protocol is used.

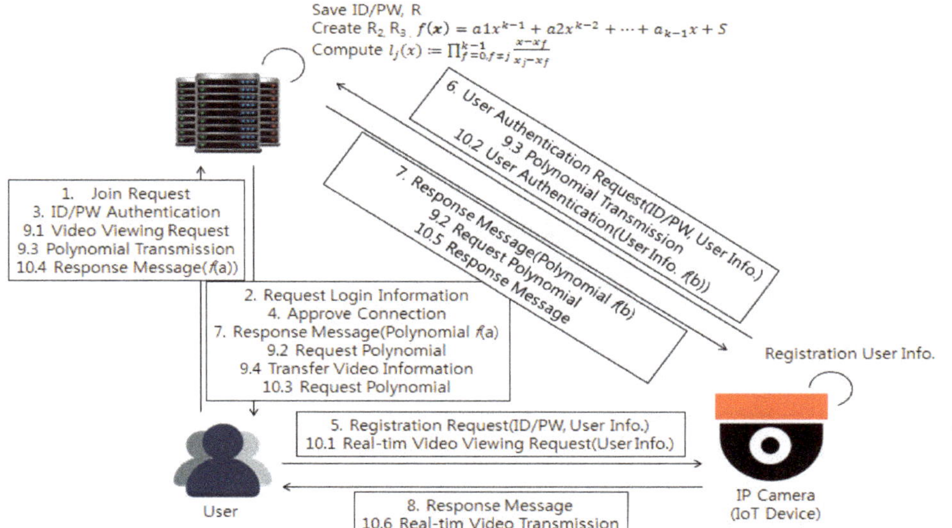

Figure 2. Protocol composition diagram.

Table 1. Proposed Notation.

Notation	Meaning
E_k(plaintext)	Encrypt a using key k
D_k(ciphertext)	Encrypt ciphertext using key k
R	Random Number
ID	Identification for authentication
PW	Password for authentication
f(k)	Polynomial for secret sharing
SN	Serial Number
l_j(x)	Formula for secret combinations

3.1. Internal Connection Protocol

This paper proposes a method of distributing keys and allowing the restoration of a key to retrieve an image only if a certain number of keys are in agreement when an image viewing request is sent internally. In the process of exchanging keys, the group key is used, and it is renewed periodically so as to secure it.

3.2. Registration Procedure

This step is a user registration procedure, and the detailed structure is shown in Figure 3.

- The user sends a joint request for registration to the control center.
- The control center sends a response message to the user, and the user proceeds to the membership subscription process based on the ID/PW, generates a random value, and transmits them together.
- The control center stores the user's subscription information in the database and finishes.
- The user sends a connection request to the IP camera through the network to register the desired IP camera.
- The IP camera requests the user's information, and the user transmits the ID/PW and the random value generated in the control center registration procedure in response to the request.
- The IP camera sends a request to the control center for the product to confirm the validity of the user.
- The control center determines the suitability of the user (i.e., the process of requesting the user to provide the serial number of the associated IP camera and the receipt of the response to the request), and sends the polynomial to the user for authentication later. The polynomial key distribution method is presented in Section 3.2.2.
- Finalizing the registration procedure of the user and the IP camera.

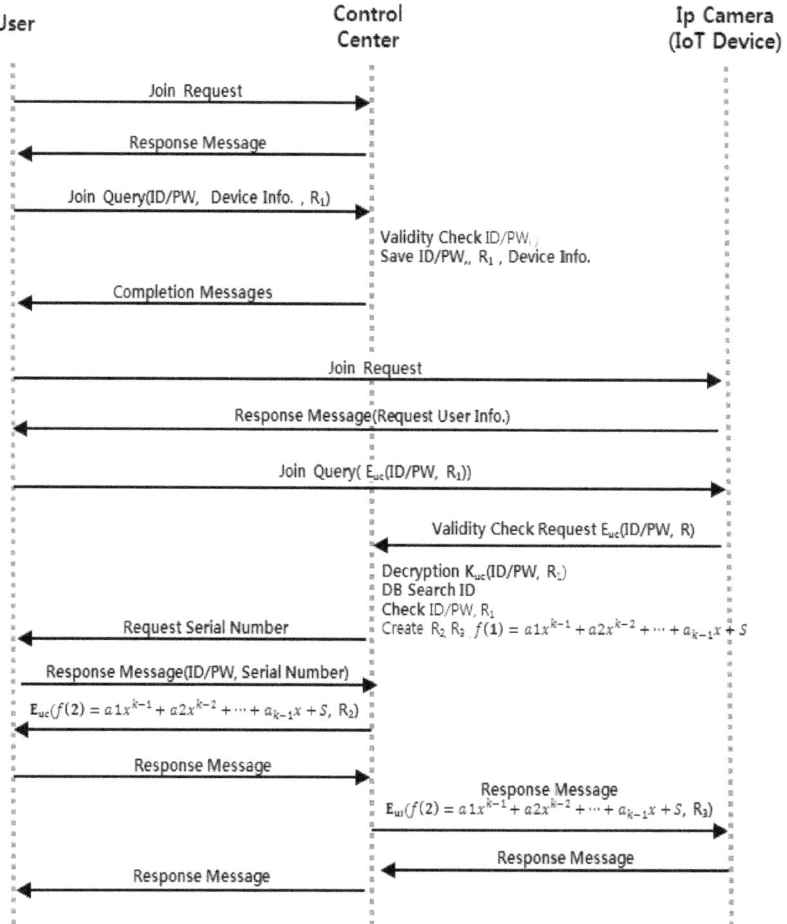

Figure 3. Registration Process Protocol.

3.2.1. Overview of Image Viewing

- The user transmits an image viewing request message.
- In this study, the user is allowed to view an image when more than an n number of devices, user and control center information, are collected. Each device determines the suitability of the user request and then transmits the polynomial information owned by the device to the operation server if it is determined to be appropriate.
- If a certain number of polynomial key values are collected, the operation server allows access to the image.

3.2.2. Polynomial Key Distribution and Transmission Method

The key distribution method uses $k - 1$ order polynomials. This paper gives an example of the simplest structure based on the three following elements: user, control center, and IP camera. Therefore, the protocol is configured by setting n to 3 and k to 2.

- The control center selects the k-1th order polynomial f(k) with the constant s.

- The control center decides the value of j and transmits f(j). In this paper, j is designated as 1 for user, 2 for control center, and 3 for IP camera, and these are encrypted using the group key transmitting.
- When the control center sends the key to the operation server, they encrypt it using the group key and transmit it.
- The control center decrypts the original key using the Lagrange polynomial if k or more distribution keys are collected.

3.2.3. Exchange of Keys to Secure Safety

An attacker may take a cipher text sent from the control center and use it as it is. In this paper, it is designed to prevent situations in which a cipher text is used as it is by using it after generating and using random values in the encryption process, and designed to mutually authenticate each other.

3.3. Video Image Monitoring

This step is a monitoring process step and the detailed structure is shown in Figure 4.

- In order to monitor a video image, the user requests to connect to the IP camera, encrypts the ID/PW and the polynomial key value using the group key, and then transmits them.
- The IP camera requests the polynomial key value of the received login information from the control center and nearby devices according to the level of security requirement.
- If the polynomial key value is confirmed to be suitable, an image is transmitted to the user in real time.

When the user ends the session, the polynomial key value is discarded, and the polynomial key value is updated according to a predetermined period.

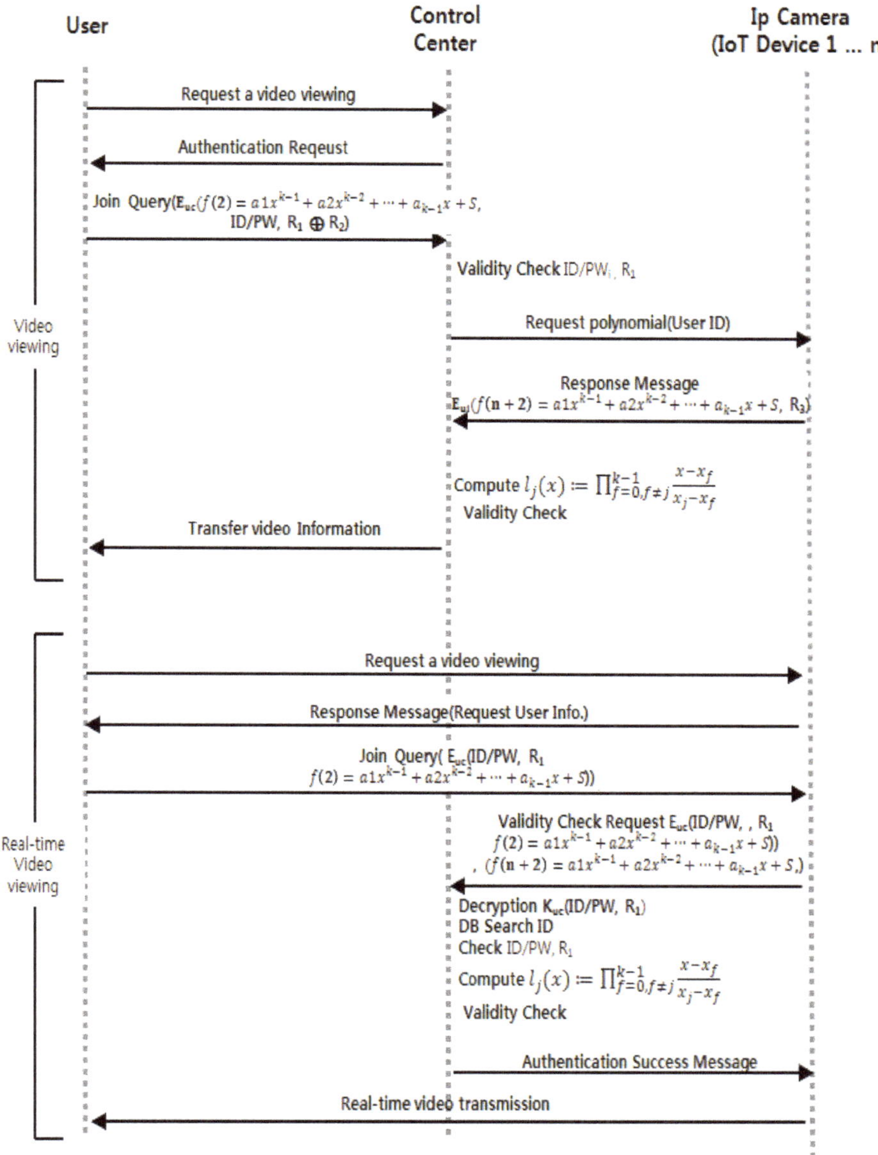

Figure 4. Monitoring process protocol.

4. Performance Evaluation

4.1. Security Evaluation

IoT environment inherits existing security threats and weaknesses of existing information and communication networks. Therefore, the performance evaluation will evaluate the safety of the proposed protocol on the security threats and additional security threats of existing information networks. The following is the result of evaluating the safety of the security threats that can occur in the information communication network.

In addition, as a result of security strength analysis, it satisfies all the security requirements defined in the OneM2M standard among the existing IoT environment standard documents, and is expected to be immediately applicable to the IoT environment.

4.1.1. Mutual Authentication

In this paper, we performed the ID/PW registration procedure using the encryption method to join the control center. In this process, it exchanges random values, which are later used to update the key and authentication values. In the control center, a polynomial f(k) is transmitted to the user and the IoT Device respectively for the subsequent authentication process. After the initial authentication, mutual authentication can be carried out by performing the verification procedure using the polynomial value. In the direct authentication process of users and IP cameras, the authentication center can perform the role of an authentication center, thus enabling secure authentication.

4.1.2. Reuse Attack

The reuse attack is a method of attack in which an unauthorized attacker can steal a message sent between each node and reuse it. To address this problem, even if a message is stolen, it is possible to authenticate it so as to prevent an attacker from using the previous transmission value by exchanging random numbers continuously. Also, in this paper, since the time stamp is assumed to be transmitted during the authentication procedure, it is possible to verify information sent the previous time.

4.1.3. Message Forgery Attack

This is a method of attack in which an unauthorized attacker seizes a message sent between each node and sends a forged message to the receiver for a desired purpose. This study confirms that data transmission is safe from message forgery attacks unless the attacker steals the key, since cipher text is generated through an encryption key before it is transmitted.

4.1.4. Sniffing

This method of attack consists in "sniffing" transmitted messages. This study confirms that it is safe from sniffing attacks because the messages between the nodes are transmitted only after applying the encryption using the inter-node secret key, which is continuously updated. Even if sniffing attempts are made in order to peek into a message, it is safe because the sniffer will only be able to see the cipher text.

4.1.5. Spoofing

This is a method of attack in which an attacker deceives the other party by changing the information that may reveal them on the network without permission. This study confirms that it is safe against spoofing attacks because the nodes have already been mutually authenticated, and the attackers will not be able to get the secret key between the nodes that have been initially shared.

4.1.6. Side-Channel Attack

This is a method of attacking through ancillary constituents such as processing time, energy usage, and electromagnetic waves. This study confirms that it is safe against side-channel attacks because it always transmits the same size message regardless of the volume of the transmitted data.

4.2. Performance Evaluation

For the performance analysis, we sequentially configured the IoT devices numbering between 3 and 40 and placed them randomly in an area measuring 45 $\times$ 45 m. We then positioned the control center in a specific location with consideration to the placement of the devices. All of the IoT devices were situated within 70 m of the control center. To confirm the efficiency regardless of the random distribution of the devices, we tested them more than 20 times under the same conditions. The following table shows the average values, while Table 2 shows the test environment in detail.

Figures 5 and 6 are simulations that compare and analyze the energy consumption of protocols in the server and client areas, respectively. Energy consumption varies depending on where the server and client are set up and is the average for more than 20 test results.

For simulated device performance, clients were based on Raspberry PI b+. In addition, for servers, PC with sufficient computational power is defined, and detailed performance is shown in Table 3. Tables 4 and 5 represent the detailed values of the simulation results.

Table 2. Assessment Environment.

Simulation Initial Settings	
Number of Device	3~40
Placement Area	45 m × 45 m
Control Center Location	X = 60 m, y = 30
Device Initial Energy	1.0
ETX, ERX	50 nanoJ
Packet Size	6000 bit

Table 3. Simulation Environment.

Sortation	Client	Server
Process	ARM1176JZF-S 700 MHz Single Core	3.5 GHz Intel Core i5-4690
Memory	512 MB	16 GB
storage medium	Micro SD Card, 8 GB	SSD 512 GB

Table 4. Compare Rates According to the Number of IoT Device (Server). (Unit: ms).

Number	ECC	RSA	SSL	Kerberos	Proposed
3	18.74498	30.77373	59.15496	0.09947	26.85049
5	31.24164	51.28955	98.59160	0.165781	44.75081
10	68.731608	112.83701	216.90153	0.364719	98.451782
20	137.46322	225.67402	433.80306	0.72944	196.90356
30	203.69549	334.40787	642.81727	1.08089	291.77528
40	281.17476	461.60595	887.32445	1.49203	402.75729

Table 5. Compare Rates According to the Number of IoT Device (Client). (Unit: ms).

Number	ECC	RSA	SSL	Kerberos	Proposed
3	12.99786	25.59363	59.44896	1.17664	0.09490
5	21.66311	42.65605	99.08161	1.96106	0.15816
10	48.95862	96.40267	223.92443	4.43201	0.35744
20	93.58461	184.27414	428.03254	8.47179	0.68325
30	134.31125	264.46751	614.30596	12.15859	0.98059
40	181.97008	358.31082	832.28549	16.47293	1.32854

In the proposed protocol, the clients perform only simple hash computation for the energy efficiency of devices with various hardware computing capabilities in the IoT environment, and the authentication protocol is proposed so that the computation is concentrated in the control center, such as encryption, decryption and polynomial computation. This feature is expected to be applicable not only to simple IP cameras but also to ultra-light devices which cannot be installed with encryption modules in the heterogeneous IoT environment, which is a limitation of lightweight security technology in the existing IoT and smart home environments.

This paper's performance analysis confirms that the proposed authentication framework is lighter than the previously well-known security authentication technology. It also confirms that it has been significantly improved in terms of security and energy efficiency due to its low energy consumption compared to the existing authentication technology. In conclusion, it is confirmed that the proposed authentication protocol is superior in terms of performance compared to the security schemes for which the existing documents are designed.

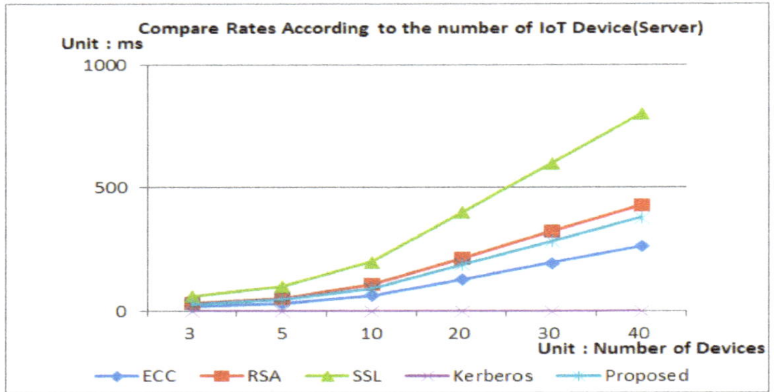

Figure 5. Energy performance evaluation(Server).

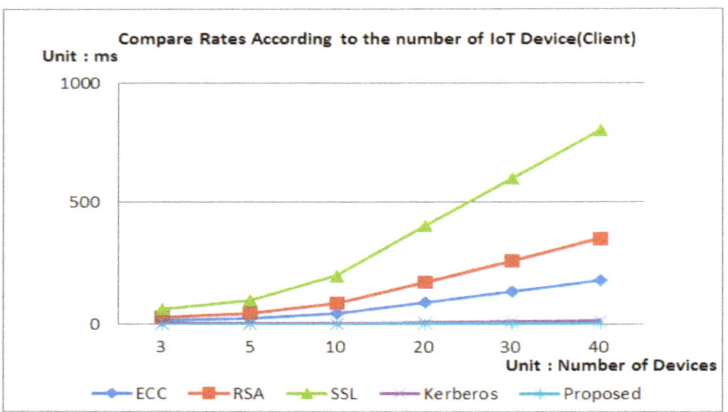

Figure 6. Energy performance evaluation(Client).

5. Conclusions

In the proposed protocol, concerning the energy efficiency of devices in IoT environments with various hardware computing abilities, only a simple hash operation was performed for the clients, while for the encryption and decryption of polynomials, the authentication protocol was proposed in order to allow the calculation to be concentrated in the control center. In addition to simple IP cameras, in a heterogeneous IoT environment, which is the limiting point of lightweight security technologies in existing IoT and smart home environments, it is expected to be utilized on all ultra-lightweight devices that cannot be equipped with encryption module.

For the authentication framework proposed through the performance analysis it was confirmed that it was lighter than the well-known security authentication technology on the client side, and in terms of devices, it can be seen that the energy consumption is significantly improved in terms of security and energy efficiency compared to the existing authentication technology.

Author Contributions: J.L. designed the protocol; J.K. researched related work; J.L. and M.-s.J. performed and analyzed the data; and J.H. and J.L. wrote the paper.

Funding: This research received no external funding.

Acknowledgments: The present Research has been conducted by the Research Grant of Kwangwoon University in 2017.

Conflicts of Interest: The authors declare no conflict of interest.

References

1. Kolias, C.; Kambourakis, G.; Stavrou, A.; Voas, J. DDoS in the IoT: Mirai and other botnets. *Computer* **2017**, *50*, 80–84. [CrossRef]
2. Yang, Y.; Wu, L.; Yin, G.; Li, L.; Zhao, H. A survey on security and privacy issues in internet-of-things. *IEEE Internet Things J.* **2017**, *4*, 1250–1258. [CrossRef]
3. Williams, C. Today the Web Was Broken by Countless Hacked Devices—Your 60-Second Summary. Available online: www.theregister.co.uk/2016/10/21/dyn_dns_ddos_explained (accessed on 21 October 2016).
4. Imperva Breaking Down Mirai: An IoT DDoS Botnet Analysis. Available online: https://www.incapsula.com/blog/malware-analysis-mirai-ddos-botnet.html (accessed on 26 October 2016).
5. Brass, I.; Tanczer, L.; Carr, M.; Blackstock, J. Regulating IoT: Enabling or Disabling the Capacity of the Internet of Things? *Risk Regul.* **2017**, *33*, 12–15.
6. Burke, D. Preventing DDOS Attacks against IoT Devices. Ph.D. Thesis, Utica College, Utica, NY, USA, 2018.
7. Frank, C.; Nance, C.; Jarocki, S.; Pauli, W.E. Protecting IoT from Mirai botnets; IoT device hardening. In Proceedings of the Conference on Information Systems Applied Research, Austin, TX, USA, 5 November 2017; p. 1508.
8. Popovic, G.; Arsic, N.; Jaksic, B.; Gara, B.; Petrovic, M. Overview, characteristics and advantages of IP Camera video surveillance systems compared to systems with other kinds of camera. *Int. J. Eng. Sci. Innov. Technol.* **2013**, *2*, 356–362.
9. Kang, J.; Han, J.; Park, J.H. Design of IP camera access control protocol by utilizing hierarchical group key. *Symmetry* **2015**, *7*, 1567–1586. [CrossRef]
10. Fularz, M.; Kraft, M.; Schmidt, A.; Kasiński, A. The architecture of an embedded smart camera for intelligent inspection and surveillance. In *Progress in Automation, Robotics and Measuring Techniques*; Springer: Cham, Switzerland, 2015; pp. 43–52.
11. Tekeoglu, A.; Tosun, A.S. Investigating security and privacy of a cloud-based wireless IP camera: NetCam. In Proceedings of the 2015 24th International Conference on Computer Communication and Networks (ICCCN), Las Vegas, NV, USA, 3–6 August 2015; pp. 1–6.
12. Bangali, J.; Shaligram, A. Design and Implementation of Security Systems for Smart Home based on GSM technology. *Int. J. Smart Home* **2013**, *7*, 201–208. [CrossRef]
13. Baran, R.; Ruść, T.; Rychlik, M. A smart camera for traffic surveillance. In *International Conference on Multimedia Communications, Services and Security*; Springer: Cham, Switzerland, 2014; pp. 1–15.
14. Krebs on Security. Researchers Find Fresh Fodder for IoT Attack Cannons. Available online: https://krebsonsecurity.com/2016/12/researchers-find-fresh-fodder-for-iot-attack-cannons/KrebsonSecurity (accessed on 16 December 2016).
15. Rio Kellyan, Tech Desk Editor. The Video of the Home Security Camera Was Hacked. Available online: https://www.bbc.com/korean/news-44962424 (accessed on 7 July 2018).
16. Korea Internet & Security Agency. 2016 Mirai Malicious Code Trends. Available online: https://www.krcert.or.kr/data/reportView.do?bulletin_writing_sequence=24864&queryString=cGFnZT0xJnNvcnRfY29kZT0mc2VhcmNoX3NvcnQ9dGl0bGVfbmFtZSZzZWFyY2hfd29yZD1taXJhaSZ4PTAmeT0w (accessed on 12 December 2016).
17. Peng, T.; Leckie, C.; Ramamohanarao, K. Survey of network-based defense mechanisms countering the DoS and DDoS problems. *ACM Comput. Surv. (CSUR)* **2007**, *39*, 3. [CrossRef]
18. Mirkovic, J.; Reiher, P. A taxonomy of DDoS attack and DDoS defense mechanisms. *ACM SIGCOMM Comput. Commun. Rev.* **2004**, *34*, 39–53. [CrossRef]

19. Jeon, W.; Kim, J.; Lee, Y.; Won, D. A practical analysis of smartphone security. In *Symposium on Human Interface*; Springer: Berlin/Heidelberg, Germany, 2011; pp. 311–320.
20. Ramachandran, V.; Nandi, S. Detecting ARP spoofing: An active technique. In Proceedings of the International Conference on Information Systems Security, Kolkata, India, 19–21 December 2005; Springer: Berlin/Heidelberg, Germany, 2005; pp. 239–250.
21. Chomsiri, T. Sniffing packets on LAN without ARP spoofing. In Proceedings of the Third 2008 International Conference on Convergence and Hybrid Information Technology, Busan, Korea, 11–13 November 2008; pp. 472–477.
22. Lin, C.F.; Yuan, S.M.; Leu, M.C.; Tsai, C.T. A framework for scalable cloud video recorder system in surveillance environment. In Proceedings of the 2012 9th international conference on Ubiquitous Intelligence & Computing and 9th International Conference on Autonomic & Trusted Computing (UIC/ATC), Fukuoka, Japan, 4–7 September 2012; pp. 655–660.
23. Lipton, A.J.; Clark, J.I.; Zhang, Z.; Venetianer, P.L.; Strat, T.; Allmen, M.; Severson, W.; Haering, N.; Chosak, A.; Frazier, M.; et al. Video Analytic Rule Detection System and Method. U.S. Patent 8,564,661, 22 October 2013.
24. Liu, H.; Chen, S.; Kubota, N. Intelligent Video Systems and Analytics: A Survey. *IEEE Trans. Ind. Inform.* **2013**, *9*, 1222–1233.
25. Lindsey, S.L.; Call, S.J. Devices, Systems, and Methods for Remote Video Retrieval. U.S. Patent Application No 14/451,067, 4 February 2016.
26. Liu, J.K.; Au, M.H.; Susilo, W.; Liang, K.; Lu, R.; Srinivasan, B. Secure sharing and searching for real-time video data in mobile cloud. *IEEE Netw.* **2015**, *29*, 46–50. [CrossRef]
27. Costin, A. Security of CCTV and video surveillance systems: Threats, vulnerabilities, attacks, and mitigations. In Proceedings of the 6th International Workshop on Trustworthy Embedded Devices, Vienna, Austria, 28 October 2016; pp. 45–54.

Article

A Scalable and Hybrid Intrusion Detection System Based on the Convolutional-LSTM Network

Muhammad Ashfaq Khan [1], Md. Rezaul Karim [2,3] and Yangwoo Kim [1,*]

1 Department of Information and Communication Engineering, Dongguk University, 30-Pildong-ro 1-gil, Jung-gu, Seoul 100-715, Korea; ashfaq_jiskani@dongguk.edu
2 Fraunhofer Institute for Applied Information Technology FIT, 53754 Sankt Augustin, Germany; rezaul.karim@fit.fraunhofer.de
3 Chair of Computer Science 5, RWTH Aachen University, 52074 Aachen, Germany
* Correspondence: ywkim@dongguk.edu; Tel.: +82-2-2260-3821

Received: 29 March 2019; Accepted: 17 April 2019; Published: 22 April 2019

Abstract: With the rapid advancements of ubiquitous information and communication technologies, a large number of trustworthy online systems and services have been deployed. However, cybersecurity threats are still mounting. An intrusion detection (ID) system can play a significant role in detecting such security threats. Thus, developing an intelligent and accurate ID system is a non-trivial research problem. Existing ID systems that are typically used in traditional network intrusion detection system often fail and cannot detect many known and new security threats, largely because those approaches are based on classical machine learning methods that provide less focus on accurate feature selection and classification. Consequently, many known signatures from the attack traffic remain unidentifiable and become latent. Furthermore, since a massive network infrastructure can produce large-scale data, these approaches often fail to handle them flexibly, hence are not scalable. To address these issues and improve the accuracy and scalability, we propose a scalable and hybrid IDS, which is based on Spark ML and the convolutional-LSTM (Conv-LSTM) network. This IDS is a two-stage ID system: the first stage employs the anomaly detection module, which is based on Spark ML. The second stage acts as a misuse detection module, which is based on the Conv-LSTM network, such that both global and local latent threat signatures can be addressed. Evaluations of several baseline models in the ISCX-UNB dataset show that our hybrid IDS can identify network misuses accurately in 97.29% of cases and outperforms state-of-the-art approaches during 10-fold cross-validation tests.

Keywords: intrusion detection system; deep learning; Spark ML; CNN; LSTM; Conv-LSTM

---

1. Introduction

Information and communication technologies now impact every aspect of society and people's lives, so attacks on ICT systems are increasing. Therefore, ICT systems need tangible, incorporated security solutions. The essential components of ICT security are confidentiality, integrity, and availability (CIA). Any activity trying to compromise CIA or avoid the security components of ICT is known as a network intrusion [1]. An intrusion detection system (IDS) is used for detecting such attacks. John et al. [2] published one of the first efforts on intrusion detection (ID) with a focus on computer security threat monitoring and surveillance. IDS is a kind of security management system utilized to observe network intrusions and nowadays is increasingly used in security systems [1,3]. An IDS typically monitors all inbound and outbound packets of a specific network to find out whether a packet shows signs of intrusion. A robust IDS can recognize the properties of maximum intrusion actions and automatically reply to them by sending warnings.

There are three main categories of IDS according to dynamic detection methods; the first is the misuse detection technique, which is known as a signature-based system (SBS). The second is the anomaly detection technique, which is known as an anomaly-based system. The third is based on a stateful protocol analysis detection approach [1,4]. The SBS depends on the pattern-matching method, comprising a signature database of identified attacks, and attempts to match these signatures with the examined data. When a match is found the alarm is raised, which is why an SBS is also known as a knowledge-based system. The misuse attack detection technique achieves maximum accuracy and minimum false alarm rate, but it cannot detect unknown attacks, while the behavior-based system is known as ABS and detects an attack by comparing abnormal behavior to normal behavior. Stateful protocol detection approaches compare the detected actions and recognize the unconventionality of the state of the protocol, and take advantage of both signature and anomaly-based attack detection approaches. In general, IDS is categorized into three types according to its architecture: Host intrusion detection system (HIDS), Network intrusion detection system (NIDS), and a hybrid approach [5,6].

A type of IDS in which a host computer plays a dynamic role in which application software is installed and useful for the monitoring and evaluation of system behavior is called a host-based intrusion detection system. In a HIDS event log files play a key role in intrusion detection [5,7]. Unlike HIDS, which evaluates every host individually, NIDS evaluates the flow of packets over the network. This kind of IDS has advantages over HIDS, which can evaluate the entire of the network with the single system, so NIDS is better in terms of computation time and the installation cost of application software on all hosts, but the foremost weakness of NIDS is its vulnerability to distribution.

A hybrid IDS, however, combines NIDS and HIDS with high flexibility and improved security mechanisms. A hybrid IDS joins the spatial sensors to address attacks that happen at a specific point or over the complete network. IDS can be classified into two main categories according to its deployment architecture: distributed and non-distributed architecture. The first kind of deployment architecture contains various ID subsystems over an extensive network, all of which communicate with each other, and is known as distributed deployment architecture; non-distributed IDS can be installed only at a single position, for example, the open-source system Snort [8]. As mentioned above, anomaly and misuse intrusion detection techniques have their limitations, but in our hybrid approach we combine the two techniques to overcome their disadvantages and propose a novel classical technique joining the benefits of the two techniques to achieve improved performance over traditional methods.

There are numerous conventional techniques for ID, for example access control mechanisms, firewalls, and encryption. These attack detection techniques have a few limitations, particularly when the systems are facing a large number of attacks like denial of service (DOS) attacks, and the systems can get a higher value of false positive and negative detection rates. In several recent studies, researchers have used machine learning (ML) techniques for intrusion detection with the ambition of improving the attack detection rates as compared to conventional attack detection techniques.

In our research, we first studied state-of-the-art approaches for IDS that apply ML techniques for ID. Then we proposed a novel approach to enhance performance in the ID domain [9]. However, simple ML techniques suffer from several limitations, while security attacks are on the increase. Upgraded learning techniques are required, particularly in features extraction and the analysis of intrusions. Hinton et al. [10] briefly explain that deep learning has achieved great success in various fields like NLP, image processing, weather prediction, etc. The techniques involved in DL have a nonlinear structure that shows a better learning capability for the analysis of composite data. The rapid progress in the parallel computing field in the last few years has also delivered a substantial hardware foundation for DL techniques.

Research has shown that a hybrid approach consisting of CNN and LSTM (aka, the Conv-LSTM network) shows a very powerful response and leads to high confidence in solving research problems such as video classification [11], sentiment [12], emotion recognition [13]; and in anomalous incident detection from a video [14]. Thus, to enhance the learning capability and detection performance of IDS, we propose a deep learning-based IDS system. In particular, we propose an improved version of IDS, which is based on Spark ML and the Conv-LSTM network. While Spark ML-based classic machine learning models help identify anomalous network traffics, the Conv-LSTM network helps identify network misuses such that both global and local latent threat signatures can be addressed. As mentioned above, ABS and SBS both have a few limitations, but if we combine the two systems we can mitigate their drawbacks. We proposed a novel IDS joining the benefits of the two systems to improve performance as compared to traditional systems. The key contributions of this research can be summarized as follows:

- We proposed an attack detection method employing IDS, which is based on Spark ML and the Conv-LSTM network. It is a novel hybrid approach, which combines both deep and shallow learning approaches to exploit their strengths and overcome analytical overheads.
- We evaluated our IDS on the ISCX-UNB dataset and analyzed the packet capture file (pcap) with Spark; earlier researchers did not consider or evaluate raw packet datasets.
- We compare our hybrid IDS with state-of-the-art IDS systems based on conventional ML. The simulation results demonstrate that our IDS can identify network misuses accurately in 97.29% of cases and outperforms state-of-the-art approaches during 10-fold cross-validation tests.
- Our proposed IDS not only outperforms existing approaches but can also achieve mass scalability while meaningfully reducing the training time, overall giving a higher degree of accuracy with a low probability of false alarms.

The rest of this article is structured as follows: background on IDS and related works are discussed in Section 2. The proposed IDS framework with architectural and implementation details is covered in Section 3. Experimental results are demonstrated in Section 4, with a comparative analysis with existing approaches. Section 5 summarizes the research and provides some possible outlooks before concluding the paper.

2. Related Work

In the last three decades, numerous anomaly detection approaches have been proposed to develop effective NIDS, aiming at good predictive accuracy to perceive attacks and upgrading the network packet traffic's speed. These approaches vary from a simple statistical learning system to classic machine learning methods and recent deep learning-based approaches. Most of these approaches attempted to extract a pattern from the network so that attack traffic can be discriminated from regular traffic.

Existing ID systems are largely based on supervised learning methods, e.g., Support vector machines (SVM) [15–17], K-nearest neighbor (KNN) [18], Random forest (RF) [19,20], etc. However, these approaches produce many false alarms and have a low detection rate for attacks in IDS. Kim et al. [21] proposed a hybrid IDS framework that integrates anomaly attack detection with misuse attack detection using the C4.decision tree (DT) classification algorithm and SVM algorithm, respectively. They evaluated their hybrid IDS on the NSL-KDD dataset. Panda et al. [22] applied the Naive Bayes (NB) algorithm for anomaly detection, which is tested on the KDD Cup dataset and found to outperform many existing IDS in terms of the low false alarm rate and low computation time with low cost. Zaman et al. [23] used an enhanced algorithm called Support Vector Decision Function (ESVDF). Their IDS was evaluated on the DARPA dataset and found to outperform other conventional techniques.

Researchers also proposed other parallel and hybrid classification approaches by amalgamating the Self-Organization Map (SOM) and the C4.classifier [24]. In this approach, the SOM-based part was envisioned to regular model behavior, and any fluctuation from that usual behavior is identified as an intrusion. The C4.classifier-based part is used for misuse detection. This approach can be used to categorize those intrusion type data into the corresponding attack category, and the final decision was made by the module known as a decision support system (DSS). The DSS was evaluated from every module by adding output and achieved maximum attack detection accuracy 99.8% on the KDD dataset along with a false alarm rate of 12.5% on the same dataset.

Albeit, the above approaches have shown good accuracy at detecting security threats to a certain degree, but it is essential to make some improvements, such as refining the accuracy and decreasing the number of false alarms [25–30]. Consequently, deep learning-based techniques are emerging, with the neural network (NN) [31] being at the core of these approaches as it provides very powerful responses not only for cybersecurity but also for other domains such as natural language processing (NLP), computer vision, and speech recognition [2,32]. Table 1 summarizes some related works.

Broadly, two fundamental characteristics account for DL-based approaches achieving tremendous success and effectiveness in these research areas: (i) hierarchical feature representations and learning capability; (ii) the ability to handle very high-dimensional data to extract valuable patterns. Previous approaches use shallow as well as deep learning techniques [32]. Gao et al. [33] proposed a restricted Boltzmann machine (RBM)-based deep belief network (DBN) to effectively learn data and identify unusual traffic from well-known datasets such as the KDD 99 dataset. Moradi et al. [34] generalized the ability of a multilayer perceptron (MLP) network analogous to layers in an attack, which is evaluated on the KDD 99 dataset. In the literature [2,35–37], LSTM-based deep learning approaches are proposed for feature selection and classification, which are evaluated on the KDD dataset.

These approaches are found to be very effective compared to the ML counterparts; researchers also proposed several ideas by combining ML- and DL-based approaches, aiming to develop robust IDS. For example, Mukkamala et al. [38] used a combined approach for classifying the connection records of the KDD 99 dataset, which is based on a support vector machine (SVM) and artificial neural network (ANN). While ANN learns the patterns of the data, the SVM is used for the classification. Javaid et al. [39] proposed a NIDS based on self-taught learning (STL). They applied their technique on the NSL-KDD dataset, and it outperformed previous approaches. Faraoun et al. [40] used multi-layered neural network backpropagation with K-means clustering for intrusion detection and experimented on the KDD 99 dataset.

On the other hand, the evolution of intrusion detection systems for the Internet of Things (IoT) is also an emerging research problem because the network traffic in real-time IoT-enabled devices is more pervasive and they are vulnerable to newer cybersecurity attacks [41]. Thus, researchers have focused on practical aspects such as mitigating the interference imposed by intruders in passive RFID networks [42].

Table 1. Overview of the state-of-the-art approaches.

Reference	Approach	Accuracy	Dataset
Yin et al. [15]	RNN IDS	90%	NSL-KDD
Reddy et al. [17]	SVM	99.95%	KDD99
B. Inger et al. [31]	ANN	99.67%	KDD99
Tsiropoulou et al. [42]	IMRA game theory	90.0%	Passive RFID
N. Gao et al. [33]	DBN	93.49%	NSL-KDD
Ghanem et al. [43]	Metaheuristic	96.4%	NSL-KDD
Sabhnani et al. [44]	MLP	97.0%	KDD99
Ying Chung et al. [45]	SSO	93.0%	KDD99
Kakavand et al. [25]	Ada boost + DT	97.0%	ISCX 2012
Kumar et al. [26]	PCA	94.05%	ISCX 2012
Yassin et al. [27]	AMGA2-NB	98.8%	ISCX 2012
Tan et al. [29]	MCA + EMD	90.12%	ISCX 2012
Sallay et al. [30]	PLL + NGL	95.30%	ISCX 2012

Note: The KDD 99 dataset contains 41 features of normal or attack types (denial of service (DOS), the user to root (U2R), remote to local (R2L), and probing attack). The NSL-KDD dataset is an improved version of the KDD 99 dataset. The ISCX 2012 dataset contains network traffic for seven days under practical and systematic circumstances.

Existing IDS systems that are typically used in traditional network intrusion detection system often fail and cannot detect many known and new security threats, largely because those approaches are based on classical ML methods that provide less focus on accurate feature selection and classification.

Consequently, many known signatures from the attack traffic remain unidentifiable and become latent. With the fast development in the field of big data and computing power, DL techniques have blossomed and been used extensively in several fields, which is why network traffic is being generated at an unprecedented scale. This imposes a great challenge to existing IDS systems because these approaches are not only unscalable but also often inefficient. To address these issues and improve the accuracy and scalability, we propose a scalable and hybrid IDS, which is based on Spark ML and the convolutional-LSTM (Conv-LSTM) network.

3. Materials and Methods

In this section, we discuss the overall architecture of the proposed approach. First, we give an overview of the architecture, which will be followed by dataset preparation. Finally, we discuss the implementation details.

3.1. Architecture of the Proposed Hybrid IDS

As shown in Figure 1, our proposed IDS system comprises of two learning stages: (i) Stage 1 is employed for anomaly detection, which is based on classic ML algorithms from the Spark ML, (ii) Stage 2 is for misuse detection, which is based on the Conv-LSTM network. To deploy such an IDS in a real-life scenario, we further incorporate the alarm module. Overall, our IDS based on this two-stage learning system will be capable of more accurate anomaly and misuse detection.

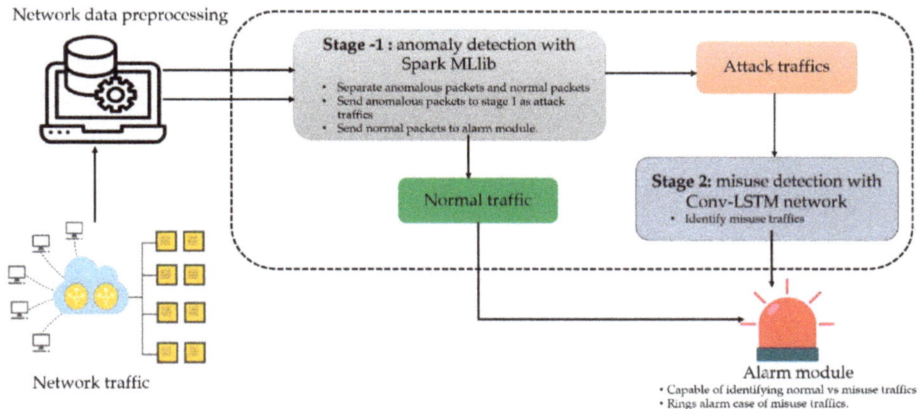

Figure 1. An overview of the proposed ID model.

3.2. Datasets

Since selecting the appropriate dataset to test a robust IDS system plays an important role, we describe and prepare the dataset before we discuss the implementation details of our proposed approach.

3.2.1. Description of the Dataset

Even though there are several benchmark ID datasets publicly available, several of them contain old-fashioned, undevitrified, inflexible, and irreproducible intrusions. To reduce these deficiencies and produce more contemporary traffic patterns, the ISCX-UNB dataset was produced by the Canadian Institute for Cybersecurity [46]. It includes several kinds of datasets to evaluate anomaly-based methods. The ISCX-IDS 2012 dataset shows realistic network behavior and comprises various intrusion scenarios. Moreover, it is shared as an entire network capture with all interior traces to evaluate payloads for deep data packet analysis.

The ISCX-IDS 2012 ID dataset contains both normal and malicious network traffic activity of seven days. The dataset was produced by profiles including abstract representations of the actions and behaviors of traffic in the network. For example, communication between the source and destination host over HTTP protocol can be denoted by packets sent and received, termination point properties, and other analogous characteristics. This representation builds a single profile. These profiles create real network traffic for HTTP, SSH, SMTP, POP3, IMAP, and FTP protocols [46].

The ISCX-IDS 2012 includes two different profiles to create network traffic behavior and scenarios. The profile that originates the anomalous or multi-stage states of attacks is known as the α profile, while the β profile characterizes features and the mathematical dissemination of the process. For instance, the β profile can contain packet size distributions in the payload specific patterns, the request of time distribution of protocol, while the α profile is constructed depending on prior attack and contains sophisticated intrusions for the individual day. There are four attack scenarios in the entire dataset according to the α profile:

- Infiltrating the network from inside
- HTTP denial of service
- Distributed denial of service using an IRC botnet
- Brute force SSH.

The complete ISCX-IDS 2012 data are summarized in Table 2. As can be seen in Table 2, every attack scenario was applied for only a single day and two days contained only regular traffic. Also, the authors of [30] explain the diversity of the regular network behavior and the complexity of the attack scenarios.

Table 2. Summary of the ISCX-IDS 2012 dataset (daily traffic).

Days	Date	Description	Size (GB)
Friday	11 June 2010	Normal, hence no malicious activity	16.1
Saturday	12 June 2010	Infiltrating the network from inside and normal activity	4.22
Sunday	13 June 2010	Infiltrating the network from inside and normal activity	3.95
Monday	14 June 2010	HTTP denial of service and normal activity	6.85
Tuesday	15 June 2010	Distributed denial of service using an IRC Botnet	23.04
Wednesday	16 June 2010	Normal, hence no malicious activity	17.6
Thursday	17 June 2010	Brute force SSH and normal activity	12.3

3.2.2. Feature Engineering and Data Preparation

As shown in Figure 1, the dump from the network traffic was initially prepared and preprocessed. The ISCX ID 2012 dataset was analyzed; after preprocessing, data were collected over seven days with the practical and systematic conditions reflecting network packet traffic and intrusions. Explicitly, the dataset is labeled for regular and malicious flows for a total of 2,381,532, and 68,792 records in each respective class. The attacks observed on the original network traffic dataset were separated into two classes, normal and malicious/abnormal.

Additionally, a variety of multi-stage attack situations were performed to produce attack traces (e.g., infiltration from the inside, HTTP, DoS, DDoS via an Internet Relay Chat (IRC) botnet, and brute force Secure Shell (SSH)). The training and test dataset distributions utilized in this study are presented in Table 3. Sections 3.3.1 and 3.3.2 give further details.

Table 3. Distribution of ISCX-IDS 2012 training and testing dataset.

Dataset/Network Flows	#Feature	Training		Testing	
ISCX-UNB Saturday	8	85,222	1353	45,889	1353
ISCX-UNB Monday	8	108,945	2451	58,664	1320
ISCX-UNB Tuesday	8	347,308	24,295	187,012	13,083
ISCX-UNB Wednesday	8	339,470	0	182,793	0
ISCX-UNB Thursday	8	255,054	3381	137,338	1822

Note: For both set: left—Benign, right—Malicious.

The essential idea was to test the consistency of the proposed novel hybrid algorithm against unknown or anomaly attack via the misuse technique. Table 4 describes detail organizations of datasets for stage-2 Conv-LSTM (Misuse) classification level for training and testing the network.

Table 4. Distribution of the data for the second stage classifier.

Input	#Features	Attack Category
Training set	8	HTTP DoS, DDoS, and Botnet
Test set	8	Brute force SSH, HTTP DoS, DDoS, Botnet, and Brute force SSH

3.3. *Implementation Details*

Since the network traffic contains both malicious and normal traffic signatures, Spark ML-based classifiers are trained to categorize the data into malicious and normal classes in stage 1. However, the Conv-LSTM network-based stage deals with malicious traffic to achieve a higher degree of ID accuracy and a low false alarm rate (FAR).

3.3.1. Stage 1: The Anomaly Detection Module

In this stage, we used the Spark ML implementation of the SVM, DT, RF, and Gradient Boosting tree (GBT) classifiers to classify attack traffic (i.e., malicious versus normal traffic). We split the training set into two subsets: 80% for the training and 20% for the testing. The classifiers were trained on the training set to learn normal vs. malicious traffic, in a binary classification setting. Then the trained classifiers are evaluated on the test set. While training these algorithms, we performed 10-fold cross-validation and grid search for the hyperparameter tuning. In each case, the best performing model was selected to evaluate the test set.

3.3.2. Stage 2: Misuse Detection and Classification Module

In this module, Conv-LSTM, used for detecting misused attacks, aims to further categorize the malicious data from stages into corresponding classification strategies, i.e., Scan, R2L, DoS, and HTTP. In LSTM the DL-based misuse attack detection technique first trained the malicious traffic to generate a model that stated the baseline profile for malicious traffic only. A schematic representation of the Conv-LSTM network is shown in Figure 2. Intuitively, an end-to-end CNN has two components: (i) a feature extractor and (ii) a classifier. The feature extractor comprises two layers called convolution and pooling layers. The extracted output, which is known as the features map, becomes the input to the second component for the classification. In this way, CNN learns the local features very well. However, the downside is that it misses the long-range interdependency of important features. Therefore, to capture the local as well as the global features more robustly, we introduced LSTM layers [47–49] after the CNN layers. In this way, we managed to address the vanishing and exploding gradient problems efficiently, which enhances the ability to ensure longer dependencies and learn efficiently from variable extent sequences [50,51].

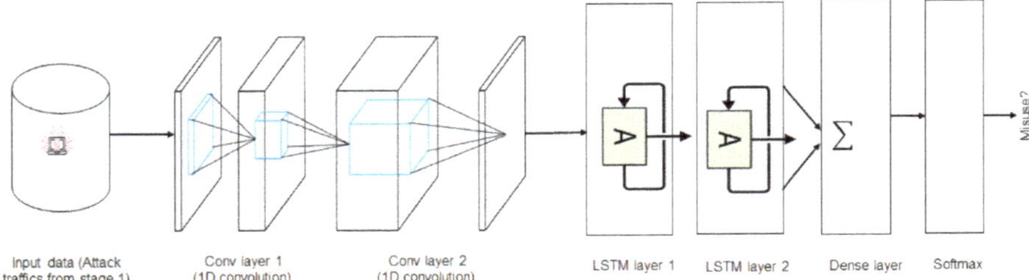

Figure 2. A schematic representation of the Conv-LSTM network, which starts by measuring attack, traffic and passing that data to both the CNN and LSTM layers before getting a flattened vector, which was fed through dense and Softmax layers for predicting the malicious traffic.

In the Conv-LSTM network, the input is initially processed by CNN, and then the output of CNN is passed through the LSTM layers to generate sequences at each time step, which helps us model both short-term and long-term temporal features [51]. Then the sequence vector is passed through a fully connected layer before feeding it into a Softmax layer for the probability distribution over the classes. In this stage, the test set is used as one of the inputs to the trained model to test if the behavior of trained traffic is normal or malicious. The attack traffic predicted by the classic models is also combined with the test set.

Then, similarly, we randomly split the dataset into training (80%) and test sets (20%) for testing. Also, 10% of the sample from the training set was used for the validation. During the training phase, first-order gradient-based optimization techniques such as Adam, AdaGrad, RMSprop, and AdaMax, with varying learning rates, were used to optimize the binary cross-entropy loss of the predicted network packet vs. the actual network packet, optimized with different combination of hyperparameters from grid search and 10-fold cross-validation to train each model on a batch size of 128. Also, we assessed the performance by adding Gaussian noise layers followed by Conv and LSTM layers to improve the model generalization and reduce overfitting.

3.4. The Alarm Module

When misuse is detected, the alarm module not only raises the alert but also compares it with normal traffic. The purpose of the alarm module is to interpret events' results on both the stage 1 and stage 2 modules. It is the last module of the proposed hybrid ID architecture that reports the ID activity to the administrator or end user.

4. Experimental Results

To show the effectiveness of our proposed hybrid approach on the ISCX ID 2012 ID dataset, we performed several experiments. We discuss the results both quantitatively and qualitatively.

4.1. Experimental Setup

The initial stage is implemented in Scala based on Spark ML. Conv-LSTM, on the other hand, was implemented in Python using Keras. Experiments were performed on a PC having a core i7 processor and 32 GB of RAM running 64-bit Ubuntu 14.04 OS. The software stack comprised of Apache Spark v2.3.0, Java (JDK) 1.8, Scala 2.11.8, and Keras. Eighty percent of the data was used for the training with 10-fold cross-validation and we evaluated the trained model based on the 20% held-over data. The Conv-LSTM is implemented in Keras and trained on an Nvidia TitanX GPU with CUDA and cuDNN, enabled to make the overall pipeline faster.

4.2. Performance Metrics

Once the models are trained, we evaluated them on the held-over test set. Then we computed the confusion matrix to compute the performance metrics. The elements of the confusion matrix help represent the predicted and expected/actual classification. The outcome of classifying is two classes: correct and incorrect. There are four fundamental situations that we considered to compute the confusion matrix:

- **True positive (TP)** measures the proportion of actual positives that are correctly identified. We specify this with x.
- **False negative (FN)** signifies the wrong predictions. More specifically, it identifies instances that are malicious but that the model incorrectly predicts as normal. We specify this with y.
- **False positive (FP)** signifies an incorrect prediction of positive, when in reality, the detected attack is normal. We specify this with z.
- **True negative (TN)** measures the proportion of actual negatives that are correctly identified attacks. We specify this with t.

Now, based on the above metrics x, y, z, and t, we have the confusion matrix in the intrusion detection setting as shown in Table 5.

Table 5. Confusion matrix for the IDS scheme.

	Predicted		
Actual	Normal	TP	FN
	Anomaly	FP	TN

From these conditions of the confusion matrix, we can calculate the performance of an IDS using the detection rate (DR) or true positive rate (TPR) and the false alarm rate (FAR), which are the two most fundamental general parameters for evaluating IDS. While DR or TPR means the ratio of intrusion instances identified by the ID model, FAR signifies the proportion of misclassified regular instances:

$$TPR = DR = TP/(TP + FN) = x/(x + y) \tag{1}$$

$$FAR = FP/(TN + FP) = z/(t + z). \tag{2}$$

When DR increases, FAR decreases. We then calculate our approach efficiency E and evaluate the hybrid IDS approach as follows:

$$E = DR/FAR. \tag{3}$$

4.3. Evaluation of the IDS System

Table 6 shows the performance of different classifiers at each stage. Only the results based on the best hyperparameters produced through an empirical random search are reported here. As shown in the table, classic model SVM performed quite poorly, giving an accuracy of only 68% in the F1-score. Tree-based classifiers managed to boost the performance significantly, showing accuracies of up to 89%.

Table 6. Performance of the classifiers at each stage.

Classifier	Precision	Recall	F1-Score	FAR	DR	Stage
SVM	0.6835	0.6515	0.6786	15.27	0.65	1
DT	0.7930	0.8012	0.7965	11.29	0.82	1
GBT	0.8529	0.8632	0.8612	8.13	0.85	1
RF	0.8919	0.8875	0.8845	5.72	0.89	1
Conv-LSTM	0.9725	0.9750	0.9729	0.71	0.97	2

However, the most significant boost that we experienced is with the Conv-LSTM network, which manages to accurately detect misuse in up to 97% of cases. The superior feature extraction of CNN and long-term dependencies between non-linear features is the reason behind this significant performance improvement. Implementation details are given in Supplementary Materials.

4.4. Overall Analysis

Table 7 compares our results with existing solutions for the ISCX-UNB dataset. This dataset was generated much later than the DARPA and KDD dataset family, so there are relatively fewer corresponding experimental results available [29]. Based on the available evaluation results for the compared methods, the best results for each study have been selected in relation to the accuracy and the false alarm rate.

Table 7. Comparison of our approach with existing solutions on the ISCX-UNB dataset.

Reference	Approach	Accuracy (DR)	False Alarm Rate
Kakavand et al. [25]	PCA	97.0	1.2
Kumar et al. [26]	AMGA2-NB	94.5	7.0
Tan et al. [29]	MCA + EMD	90.12	7.92
Sally et al. [30]	PLL + NGL	95.31	0.80
Our approach	Spark ML + Conv-LSTM	97.29	0.71

It can be observed that our proposed system performs better both in relation to the accuracy and the false alarm rate associated with advanced techniques, mainly because of the efficient feature selection technique we used and the implementation of a suitable Spark ML and Conv-LSTM approach. It is worth noting that these comparisons are for reference only as many researchers have used different proportions of traffic types and dataset distributions, preprocessing techniques, and sampling methods.

Therefore, a straightforward comparison of some metrics, such as training and testing time, is usually not considered appropriate, although our hybrid approach achieved improved performance in terms of all the performance metrics and outperformed other approaches. Nevertheless, we state that one can achieve a remarkable level of security against intrusion attacks using the hybrid technique, which is simple, fast, vigorous, and highly appropriate for real-time applications as well.

5. Conclusions and Outlook

In this paper, a hybrid IDS is proposed, implemented, and evaluated on the well-known ISCX-UNB dataset. The presented IDS is largely based on Spark ML and the Conv-LSTM network, which is particularly useful for the cybersecurity research. Our proposed IDS, which is a two-stage learning platform, shows good predictive accuracy at each stage, which is not only the case for the classification algorithms such as DT, RF, GBT, and SVM but also for the Conv-LSTM network, giving 97.29% predictive accuracy.

The proposed hybrid IDS based on the Conv-LSTM network integrated the power of both CNN and the LSTM network, which can be thought of as an efficient approach having both AB (Anomaly-based) and SB (Signature-based) classification approaches. The modular and hierarchical structure of our IDS not only performs better than state-of-the-art IDS approaches in terms of DR and accuracy for intrusion detection but also reduces the computational complexity.

However, one possible downside of our approach is that we have tested our IDS on only a single dataset. It is important to test it on a more recent dataset since the signature of the attack traffic often changes. In the future, we intend to: (i) extend our work so that anomaly and network misuses can be detected on real-time streaming data, and (ii) focus on exploring DL as an attribute extraction tool to learn competent data illustrations in case of other anomaly recognition problems in a more recent dataset.

Supplementary Materials: The source codes of the implementation are available on GitHub at https://github.com/rezacsedu/Intrusion-Detection-Spark-Deep-Learning.

Author Contributions: M.A.K. conceived the research, wrote the paper, designed the framework, and performed the experiments. M.R.K. developed the algorithm, helped with the experiments, and contributed to the background research. Y.K. and M.R.K. assisted with the proofreading, revision, and improvements. Y.K. supervised the overall research.

Funding: This research was supported by the MSIT (Ministry of Science, ICT), Korea, under the ITRC (Information Technology Research Center) support program (IITP-2018-2016-0-00465) supervised by the IITP (Institute for Information & communications Technology Promotion).

Conflicts of Interest: The authors declare no conflict of interest.

Abbreviations

RNN	Recurrent Neural Network
Conv-LSTM	Convolutional-Long short-term memory
IDS	Intrusion detection system
DL	Deep learning
ICT	Information and communication technology
CIA	Confidentiality, integrity, and availability
SBS	Signature-based system
HIDS	Host intrusion detection system
NIDS	Network intrusion detection system
DoS	Denial of service
U2R	User to root
R2L	Remote to local STL
STL	Self-taught learning
NN	Neural network
MLP	Multilayer perceptron
SVM	Support vector machine
GBT	Gradient Boosting tree
DBN	Deep belief network
DSS	Decision support system
DT	Decision tree
SSO	Simplified swarm optimization
NB	Naive Bayes
ESVDF	Enhanced Support Vector Decision Function

References

1. Xu, C.; Shen, J.; Du, X.; Zhang, F. An intrusion detection system using a deep neural network with gated recurrent units. *IEEE Access* **2018**, *6*, 48697–48707. [CrossRef]
2. Vinayakumar, R.; Soman, K.P.; Poornachandran, P. Applying convolutional neural network for network intrusion detection. In Proceedings of the International Conference on Advances in Computing, Communications and Informatics (ICACCI), Udupi, India, 13–16 September 2017; pp. 1222–1228.
3. Sharma, S.; Gupta, R. Intrusion detection system: A review. *Int. J. Secur. Its Appl.* **2015**, *9*, 69–76. [CrossRef]
4. Allen, J.; Christie, A.; Fithen, W.; Mchugh, J.; Pickel, J. *State of the Practice of Intrusion Detection Technologies*; Carnegie-Mellon Univ Pittsburgh Pa Software Engineering Inst: Pittsburgh, PA, USA, 2000.
5. Mighan, S.N.; Kahani, M. Deep Learning Based Latent Feature Extraction for Intrusion Detection. In Proceedings of the Iranian Conference on Electrical Engineering (ICEE), Mashhad, Iran, 8–10 May 2018; pp. 1511–1516.
6. Bijone, M. A survey on secure network: Intrusion detection prevention approaches. *Am. J. Inf. Syst.* **2016**, *4*, 69–88.
7. Hodo, E.; Bellekens, X.; Hamilton, A.; Tachtatzis, C.; Atkinson, R. Shallow and deep networks intrusion detection system: A taxonomy and survey. *arXiv* **2017**, arXiv:1701.02145.

8. Axelsson, S. *Intrusion Detection Systems: A Survey and Taxonomy*; Technical Report; Chalmers University: Goteborg, Sweden, 2000; Volume 99.
9. Kim, J.; Kim, H. An effective intrusion detection classifier using long short-term memory with gradient descent optimization. In Proceedings of the IIEEE international Conference on Platform Technology and Service (PlatCon), Busan, Korea, 13–15 February 2017.
10. Hinton, G.E.; Osindero, S.; Teh, Y.W. A fast learning algorithm for deep belief nets. *Neural Comput.* **2006**, *18*, 1527–1554. [CrossRef]
11. Wu, Z.; Wang, X.; Jiang, Y.G.; Ye, H.; Xue, X. Modeling spatial-temporal clues in a hybrid deep learning framework for video classification. In Proceedings of the 23rd ACM International Conference on Multimedia, Brisbane, Australia, 26–30 October 2015; pp. 461–470.
12. Tang, D.; Qin, B.; Liu, T. Document modeling with gated recurrent neural network for sentiment classification. In Proceedings of the Conference on Empirical Methods in Natural Language Processing (EMNLP), Lisbon, Portugal, 17–21 September 2015; pp. 1422–1432.
13. Fan, Y.; Lu, X.; Li, D.; Liu, Y. Video-based emotion recognition using CNN-RNN and C3D hybrid networks. In Proceedings of the 18th ACM International Conference on Multimodal Interaction, Tokyo, Japan, 12–16 November 2016; pp. 445–450.
14. Vignesh, K.; Yadav, G.; Sethi, A. Abnormal Event Detection on BMTT-PETS Surveillance Challenge. In Proceedings of the IEEE Conference on Computer Vision and Pattern Recognition Workshops (CVPRW), Honolulu, HI, USA, 21–26 July 2017; pp. 2161–2168.
15. Yin, C.; Zhu, Y.; Fei, J.; He, X. A Deep Learning Approach for Intrusion Detection Using Recurrent Neural Networks. *IEEE Access* **2017**, *5*, 21954–21961. [CrossRef]
16. Kuang, F.; Xu, W.; Zhang, S. A novel hybrid KPCA and SVM with GA model for intrusion detection. *Appl. Soft Comput.* **2014**, *18*, 178–184. [CrossRef]
17. Reddy, R.R.; Ramadevi, Y.; Sunitha, K.V.N. Effective discriminant function for intrusion detection using SVM. In Proceedings of the International Conference on Advances in Computing, Communications and Informatics (ICACCI), Jaipur, India, 21–24 Septembert 2016; pp. 1148–1153.
18. Li, W.; Yi, P.; Wu, Y.; Pan, L.; Li, J. A new intrusion detection system based on KNN classification algorithm in wireless sensor network. *J. Electr. Comput. Eng.* **2014**, *2014*, 240217. [CrossRef]
19. Farnaaz, N.; Jabbar, M.A. Random forest modeling for network intrusion detection system. *Procedia Comput. Sci.* **2016**, *89*, 213–217. [CrossRef]
20. Zhang, J.; Zulkernine, M.; Haque, A. Random-Forests-Based Network Intrusion Detection Systems. *IEEE Trans. Syst. Man Cybern. Part C Appl. Rev.* **2008**, *38*, 649–659. [CrossRef]
21. Kim, G.; Lee, S.; Kim, S. A novel hybrid intrusion detection method integrating anomaly detection with misuse detection. *Expert Syst. Appl.* **2014**, *41*, 1690–1700. [CrossRef]
22. Panda, M.; Patra, M.R. Network intrusion detection using naive bays. *Int. J. Comput. Sci. Netw. Secur.* **2007**, *7*, 258–263.
23. Zaman, S.; Karray, F. Features selection for intrusion detection systems based on support vector machines. In Proceedings of the IEEE Consumer Communications and Networking Conference, Las Vegas, NV, USA, 10–13 January 2009; pp. 1–8.
24. Depren, O.; Topallar, M.; Anarim, E.; Ciliz, M.K. An intelligent intrusion detection system (IDS) for anomaly and misuse detection in computer networks. *Expert Syst. Appl.* **2005**, *4*, 713–722. [CrossRef]
25. Kakavand, M.; Mustapha, N.; Mustapha, A.; Abdullah, M.T. Effective Dimensionality Reduction of Payload-Based Anomaly Detection in TMAD Model for HTTP Payload. *KSII Trans. Internet Inf. Syst.* **2016**, *10*, 3884–3910.
26. Kumar, G.; Kumar, K. Design of an evolutionary approach for intrusion detection. *Sci. World J.* **2013**, *2013*, 962185. [CrossRef]
27. Yassin, W.; Udzir, N.I.; Muda, Z.; Sulaiman, M.N. Anomaly-based intrusion detection through k-means clustering and naives Bayes classification. In Proceedings of the 4th International Conference on Computing and Informatics, ICOCI, Kuching, Malaysia, 28–30 August 2013; Volume 49, pp. 298–303.
28. Tahir, H.M.; Said, A.M.; Osman, N.H.; Zakaria, N.H.; Sabri, P.N.A.M.; Katuk, N. Oving K-means clustering using discretization technique in network intrusion detection system. In Proceedings of the 3rd International Conference on Computer and Information Sciences (ICCOINS), Kuala Lumpur, Malaysia, 15–17 August 2016; pp. 248–252.

29. Tan, Z.; Jamdagni, A.; He, X.; Nanda, P.; Liu, R.P.; Hu, J. Detection of Denial-of-Service Attacks Based on Computer Vision Techniques. *IEEE Trans. Comput.* **2015**, *64*, 2519–2533. [CrossRef]
30. Sallay, H.; Ammar, A.; Saad, M.B.; Bourouis, S. A real time adaptive intrusion detection alert classifier for high speed networks. In Proceedings of the IEEE 12th International Symposium on Network Computing and Applications (NCA), Cambridge, MA, USA, 22–24 August 2013; pp. 73–80.
31. Ingre, B.; Yadav, A. Performance analysis of NSL-KDD dataset using ANN. In Proceedings of the IEEE International Conference on Signal Processing and Communication Engineering Systems, Guntur, India, 2–3 January 2015; pp. 92–96.
32. Lipton, Z.C.; Berkowitz, J.; Elkan, C. A critical review of recurrent neural networks for sequence learning. *arXiv* **2015**, arXiv:1506.00019.
33. Gao, N.; Gao, L.; Gao, Q.; Wang, H. An intrusion detection model based on deep belief networks. In Proceedings of the IEEE Second International Conference on Advanced Cloud and Big Data, Huangshan, China, 20–22 November 2014; pp. 247–252.
34. Moradi, M.; Zulkernine, M. A neural network-based system for intrusion detection and classification of attacks. In Proceedings of the IEEE International Conference on Advances in Intelligent Systems-Theory and Applications, Guwahati, India, 4–6 March 2004; pp. 15–18.
35. Staudemeyer, R.C.; Omlin, C.W. Extracting salient features for network intrusion detection using machine learning methods. *S. Afr. Comput. J.* **2014**, *52*, 82–96. [CrossRef]
36. Staudemeyer, R.C.; Omlin, C.W. Evaluating performance of long short-term memory recurrent neural networks on intrusion detection data. In Proceedings of the South African Institute for Computer Scientists and Information Technologists Conference, East London, Africa, 7–9 October 2013; pp. 218–224.
37. Staudemeyer, R.C. Applying long short-term memory recurrent neural networks to intrusion detection. *S. Afr. Comput. J.* **2015**, *56*, 136–154. [CrossRef]
38. Mukkamala, S.; Sung, A.H.; Abraham, A. Intrusion detection using an ensemble of intelligent paradigms. *J. Netw. Comput. Appl.* **2005**, *28*, 167–182. [CrossRef]
39. Javaid, A.; Niyaz, Q.; Sun, W.; Alam, M. A deep learning approach for network intrusion detection system. In Proceedings of the 9th EAI International Conference on Bio-inspired Information and Communications Technologies (formerly BIONETICS), ICST (Institute for Computer Sciences, Social-Informatics and Telecommunications Engineering), New York, NY, USA, 3–5 December 2016; pp. 21–26.
40. Faraoun, K.M.; Boukelif, A. Neural networks learning improvement using the K-means clustering algorithm to detect network intrusions. *INFOCOMP* **2006**, *5*, 28–36.
41. Santos, L.; Rabadao, C.; Gonçalves, R. Intrusion detection systems in Internet of Things: A literature review. In Proceedings of the IEEE 13th Iberian Conference on Information Systems and Technologies (CISTI), Caceres, Spain, 13–16 June 2018; pp. 1–7.
42. Tsiropoulou, E.E.; Baras, J.S.; Papavassiliou, S.; Qu, G. On the Mitigation of Interference Imposed by Intruders in Passive RFID Networks. In *International Conference on Decision and Game Theory for Security*; Springer: Cham, Switzerland, 2016; pp. 62–80.
43. Ghanem, T.F.; Elkilani, W.S.; Abdul-Kader, H.M. A hybrid approach for efficient anomaly detection using metaheuristic methods. *J. Adv. Res.* **2015**, *6*, 609–619. [CrossRef]
44. Sabhnani, M.; Serpen, G. Application of Machine Learning Algorithms to KDD Intrusion Detection Dataset within Misuse Detection Context. In Proceedings of the International Conference on Machine Learning: Models, Technologies, and Applications (MLMTA), Las Vegas, NV, USA, 23–26 June 2003; pp. 209–215.
45. Chung, Y.Y.; Wahid, N. A hybrid network intrusion detection system using simplified swarm optimization (SSO). *Appl. Soft Comput.* **2012**, *12*, 3014–3022. [CrossRef]
46. Shiravi, A.; Shiravi, H.; Tavallaee, M.; Ghorbani, A.A. Toward developing a systematic approach to generate benchmark datasets for intrusion detection. *Comput. Secur.* **2012**, *31*, 357–374. [CrossRef]
47. Hochreiter, S.; Schmidhuber, J. Long Short-Term Memory. Long short-term memory. *Neural Comput.* **1997**, *9*, 1735–1780. [CrossRef]
48. Gers, F.A.; Schraudolph, N.N.; Schmidhuber, J. Learning precise timing with LSTM recurrent networks. *J. Mach. Learn. Res.* **2002**, *3*, 115–143.
49. Giancarlo, Z.; Karim, M.R. *Deep Learning with TensorFlow: Explore Neural Networks and Build Intelligent Systems with Python*; Packt Publishing Ltd.: Birmingham, UK, 2018.

50. Khan, M.A.; Karim, M.R.; Kim, Y. A Two-Stage Big Data Analytics Framework with Real World Applications Using Spark Machine Learning and Long Short-Term Memory Network. *Symmetry* **2018**, *10*, 485. [CrossRef]
51. Karim, M.R.; Cochez, M.; Dietrich-Rebholz, S. Recurrent Deep Embedding Networks for Genotype Clustering and Ethnicity Prediction. *arXiv* **2018**, arXiv:1805.12218.

Article

Battlefield Target Aggregation Behavior Recognition Model Based on Multi-Scale Feature Fusion

Haiyang Jiang, Yaozong Pan, Jian Zhang and Haitao Yang *

Space Engineering University, 81 Road, Huairou District, Beijing 101400, China; wzc@bupt.deu.cn (H.J.); yfy@bupt.deu.cn (Y.P.); zhj@bupt.deu.cn (J.Z.)

* Correspondence: 13400416091@sjtu.edu.cn

Received: 15 May 2019; Accepted: 31 May 2019; Published: 5 June 2019

Abstract: In this paper, our goal is to improve the recognition accuracy of battlefield target aggregation behavior while maintaining the low computational cost of spatio-temporal depth neural networks. To this end, we propose a novel 3D-CNN (3D Convolutional Neural Networks) model, which extends the idea of multi-scale feature fusion to the spatio-temporal domain, and enhances the feature extraction ability of the network by combining feature maps of different convolutional layers. In order to reduce the computational complexity of the network, we further improved the multi-fiber network, and finally established an architecture—3D convolution Two-Stream model based on multi-scale feature fusion. Extensive experimental results on the simulation data show that our network significantly boosts the efficiency of existing convolutional neural networks in the aggregation behavior recognition, achieving the most advanced performance on the dataset constructed in this paper.

Keywords: machine vision; aggregation behavior; convolutional neural network; video; action recognition

---

1. Introduction

Battlefield target aggregation behavior is a common group behavior in the joint operations environment, which is usually a precursor to important operational events such as force adjustment, battle assembly, and sudden attack. To grasp the battlefield initiative, it is important to identify the aggregation behavior of enemy targets. The intelligence video records the different behaviors of the battlefield targets, and effectively identifying the aggregate behavior in the video is the main purpose of this paper.

For the time being, the identification of battlefield aggregation behavior requires a manual interpretation, which is inefficient in battlefield environments. It is an inevi trend for intelligent battlefield development to introduce intelligent recognition algorithms to identify the aggregation behavior. For behavior recognition, intelligent algorithms based on deep learning are the research hotspots. In particular, 3D Convolutional Neural Networks (3D-CNN), which show significant results in behavior recognition, provide a technical basis for battlefield target aggregation behavior recognition.

Unfortunately, the traditional 3D-CNN model has certain drawbacks for the battlefield target aggregation behavior: (1) Compared with the human behavior in the video, the proportion of the target is uncertain in the intelligence video. The existing 3D-CNN network lacks the interaction of multi-scale features. The loss of spatial information in the down-sampling process has a great influence on the detection rate of aggregated behavior; (2) The duration of the aggregation behavior is uncertain, and the down-sampling in the temporal dimension will cause the loss of timing information, which will indirectly affect the final recognition accuracy; (3) Traditional 3D-CNN are computationally expensive. Therefore, the network structure is difficult to flexibly expand, and it is also difficult to cope with large-scale identification tasks.

Our article does not consider the disturbances of complex environmental factors (such as complex weather, etc.), and only focuses on solving aggregation behavior recognition problems with deep learning networks.

We have improved the traditional 3D-CNN. On the one hand, we construct a multi-scale feature fusion 3D-CNN model, which combines multi-scale spatio-temporal data of different convolutional layers to promote the interaction between multi-scale information. This model effectively reduces the information loss caused by the network down-sampling. The model proposed in this paper effectively solves the problem that the size of the battlefield target is different and that the aggregation behavior duration is uncertain, which effectively improves the final recognition accuracy. On the other hand, this paper uses the improved spatio-temporal multi-fiber network as the backbone network, which slices a complex neural network into an ensemble of lightweight networks or fibers. Our network effectively overcomes the huge computational problem of 3D-CNN. At the same time, the depth of the network is deepened, and the nonlinear expression ability of the neural network is increased.

The rest of the paper is organized as follows. In Section 2, related work is discussed. We present our method in Section 3 and the dataset in Section 4. We report the experimental results in Section 5. The conclusion is in Section 6.

2. Related Work

At present, human behavior recognition is a research hotspot in the field of intelligent video analysis. Battlefield target aggregation behavior recognition and human behavior recognition are in the same field of behavior recognition but are not identical. The main reason for this is that single-frame information, which contributes a lot to the human behavior recognition, contributes less to the aggregation behavior recognition. Therefore, multi-frame information must be processed in order to enhance the recognition effect of aggregation behavior. In recent years, researchers have proposed a number of methods for video behavior recognition, which are mainly divided into traditional feature extraction methods [1–3] and the method based on deep learning [4,5].

The early traditional methods based themselves on the description of spatio-temporal interest points to extract the features in the video. Wang proposed a dense trajectory method [6], which extracts local features along trajectories guided by an optical flow. This method achieves a state-of-the-art level in the traditional method. However, the extraction process of the traditional underlying features is independent of the specific tasks, and the wrong feature selection will bring great difficulties to the identification. In addition, due to the cumbersome feature calculation, the traditional method is gradually replaced by deep learning.

With the wide application of deep learning in the image field, the video behavior recognition method based on deep learning has gradually become a new hotspot in the field of behavior recognition. The two-stream architecture [7] uses RGB (Red-Green-Blue) frames and optical flows between adjacent frames as two separate inputs of the network, and fuses their output classification scores as the final prediction. Wang [8] constructs a long-term domain structure based on a two-stream architecture. First, multiple video segments are extracted by sparse sampling, and then a two-stream convolution network is established on each segment. Finally, the output results of all the networks are combined for the prediction classification. Many works follow a two-stream architecture and extend this architecture [9–11]. RNN (Recursive Neural Network) is excellent in capturing the timing information. Inspired by it, Donahue [12] combines CNN (Convolutional Neural Networks) and RNN to propose a long-term recursive convolutional neural network. More recently, with the increasing computing capability of modern GPUs, 3D-CNN has drawn more and more attention. Varol [13] designed a long-term convolutional network by extending the length of input of the 3D convolutional network, and studied the influence of different inputs on the recognition results. Carreira [14] introduced the two-stream idea into 3D-CNN, innovatively used ImageNet to pre-train 2D-CNN, and then replicated the parameters of the 2D convolution kernel in the time dimension to form a 3D convolution kernel. The recognition accuracy was significantly improved. Although 3D-CNN can learn the motion characteristics from the original frame end-to-end, the network parameters and calculation amount are huge, so the experimental training and testing need to occupy huge resources. Qiu [15] proposed Pseudo-3D (P3D), which decomposes a 3D convolution of $3 \times 3 \times 3$ into a 2D convolution of $1 \times 3 \times 3$, followed by a 1D convolution of $3 \times 1 \times 1$. In addition, S3D [16] and R(2+1)D [17] also applied a similar architecture. Multi-fiber networks [18], which use multi-frame RGB as the input, greatly reduce the computational complexity on the basis of ensuring a recognition accuracy. FPN [19] combines down-top, top-down and lateral connections with using high and low semantic features, which improves the recognition accuracy.

Traditional 3D networks lack the use of multi-scale information, which affects the recognition accuracy of the network. The traditional two-stream model uses 2D-CNN to process images, which is better than the single-stream model but lacks the ability to extract temporal information. In view of the above problems, this paper combines the advantages of 3D-CNN and the two-stream network structure to construct a 3D convolution. The two-stream model is based on multi-scale feature fusion. The experimental results show that the model has a high efficiency and accuracy.

3. The Proposed Method

Battlefield target aggregation is a kind of behavior which can be regarded as the process of the combat units gathering from the starting position to the target, with obvious temporal and spatial characteristics. In this paper, we have proposed a new ConvNet called 3D convolution Two-Stream model based on multi-scale feature fusion. As shown in Figure 1, the 3D ConvNets based on multi-scale feature fusion (M3D) extracts the features of RGB sequences and optical flow sequences respectively, and obtains recognition results by averaging the output of the two networks.

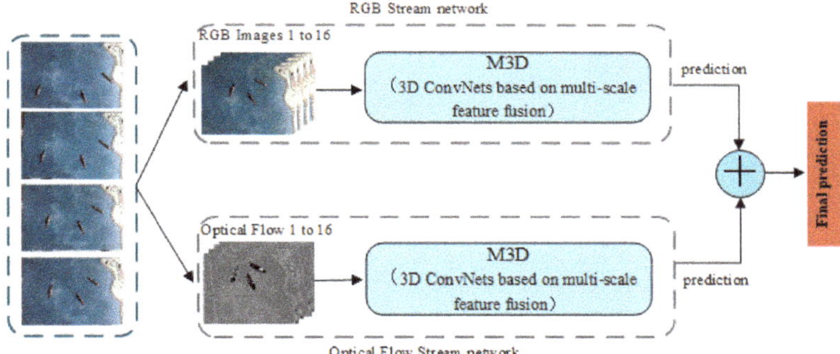

Figure 1. Two-Stream M3D (3D convolution Two-Stream model based on multi-scale feature fusion). M3D (3D ConvNets based on multi-scale feature fusion) extracts the features of RGB (Red-Green-Blue) sequences and optical flow sequences respectively, and gets the recognition results by averaging the output of the two networks.

The RGB network is a deep learning network that performs behavior recognition by extracting the spatio-temporal information of multi-frame RGB images.

The input of the optical flow network is a sequence of optical flow images that contains motion information. Although RGB images can provide rich exterior information, they are likely to cause background interference. Optical flow images can shield background interference, which helps the network understand the motion information of the target. Therefore, the optical flow network can effectively pay attention to the motion information in the video. The backbone network of this paper adopts a modular design, and the main body is composed of multi-fiber modules.

3.1. Multi-Fiber Unit (MF Unit)

In order to reduce the computational cost and increase the depth of the network, we use the improved spatio-temporal multi-fiber network as the backbone network. The calculation amount of the network is closely related to the number of connections between the two layers. The traditional conventional unit uses two convolutional layers to learn features, which is straightforward but computationally expensive. The total number of connections between these two layers can be computed as:

$$C = (M_{in} \times M_{mid} + M_{mid} \times M_{out}), \tag{1}$$

where C represents the connections, M_{in} represents the number of input channels, M_{mid} represents the number of middle channels and M_{out} represents the number of output channels.

Equation (1) indicates that the number of network connections is the quadratic of the network width.

The spatio-temporal multi-fiber convolution network is modular in design, with the multi-fiber unit splitting a single path into N parallel paths, each path being isolated from the other paths. As shown in Equation (2), the total width of the unit remains the same, but the number of connections is reduced to the original $1/N$. In this paper, we set $N = 16$.

$$C = N \times (\frac{M_{in}}{N} \times \frac{M_{mid}}{N} + \frac{M_{mid}}{N} \times \frac{M_{out}}{N}) = \frac{(M_{in} \times M_{mid} + M_{mid} \times M_{out})}{N} \tag{2}$$

The multiplexer can share information between N paths. The first $1 \times 1 \times 1$ convolutional layer is responsible for merging the features and reducing the number of channels, and the second $1 \times 1 \times 1$ convolution layer distributes the feature map to each channel. The parameters within the multiplexer are randomly initialized and automatically adjusted by end-to-end backpropagation. The multi-fiber module and multiplexer are shown in Figure 2.

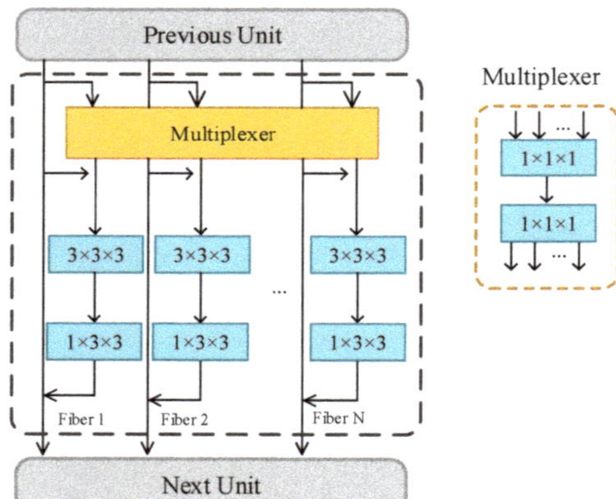

Figure 2. Multi-fiber unit and multiplexer. The multiplexer module was incorporated to facilitate the information flow between the fibers.

3.1.1. 3D ConvNets Based on Multi-Scale Feature Fusion (M3D)

As shown in Figure 3, 3D ConvNets based on multi-scale feature fusion (M3D) consists of a mainstream network and two tributary networks. The input size of the mainstream network is 16 × 224 pixels × 224 pixels. The network settings are shown in Table 1. We carry out the spatio-temporal down-sampling in Conv5_1, Conv3_1 and Conv4_1 with stride (2, 2, 2). In Conv1 and MaxPool, the down-sampling of the spatio was carried out, and the stride is (1, 2, 2). The output of Conv5_3 is the averaged spatio-temporal pooling, and results in all kinds of recognition probabilities through the fully connected layer.

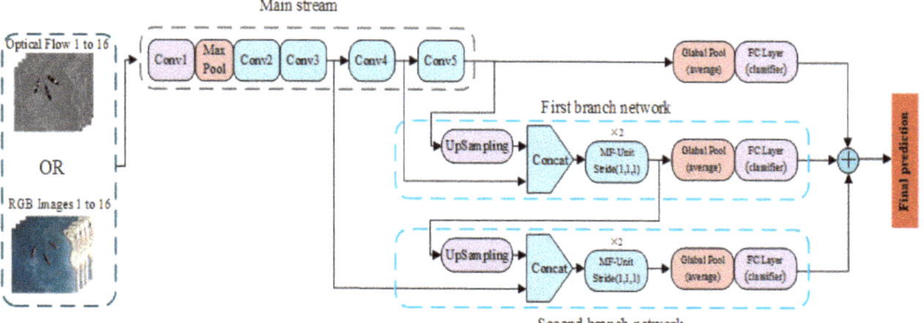

Figure 3. Network architecture details for M3D. The input of the first branch is the output of Conv5_3 and Conv4_6. The input of the second branch is the output of the first branch convolution module and the output of Conv3_4. The classification probability of the three-way network is averaged to obtain the final recognition probability. In the first branch network, in order to splice the output feature maps of Conv5_3 and Conv4_6, we use the up-sampling layer to up-sample the feature map of Conv5_3 from $4 \times 7 \times 7$ to $8 \times 14 \times 14$. The spliced feature map is sent to the convolution module to extract the features, and after the pooling layer the full connection layer obtains the classification probability of the branch. In the second branch network, with the first branch network feature map, we up-sample the spatio-temporal resolution by a factor of 2 (using nearest neighbor up-sampling for simplicity). The up-sampled map is then merged with the feature map of Conv3_4. The spliced feature map is sent to the convolution module to extract the features and results in the classification probability after the fully connected layer. Finally, the classification probability of the three-way network is averaged to obtain the final recognition probability.

Table 1. Mainstream network settings. When the input is an optical flow frame, the number of input channels of the network is 2. When the input is an RGB image, the network input channel is 3. The stride is denoted by "(temporal stride, height stride, and width stride)".

Layer	Repeat	Channel	Stride	Output Size
Input		3(RGB)/2(Flow)		$16 \times 224 \times 224$
Conv1	1	16	(1, 2, 2)	$16 \times 112 \times 112$
MaxPool		16	(1, 2, 2)	$16 \times 56 \times 56$
Conv2_X (MF Unit)	3	96	(1, 1, 1)	$16 \times 56 \times 56$
Conv3_X (MF Unit)	1 3	192	(2, 2, 2) (1, 1, 1)	$8 \times 28 \times 28$
Conv4_X (MF Unit)	1 5	384	(2, 2, 2) (1, 1, 1)	$4 \times 14 \times 14$
Conv5_X (MF Unit)	1 2	768	(2, 2, 2) (1, 1, 1)	$2 \times 7 \times 7$
AvgPooling				$1 \times 1 \times 1$
FC				2

In the process of multi-scale feature map fusion, it must be ensured that the multi-scale feature can retain the original information after fusion and maintain the validity of the fusion feature. In view of the above situation, this paper proposes a number of fusion methods, and the function is expressed as:

$$y = f\left(x^a, x^b\right), \tag{3}$$

where $x^a \in P^{t \times h \times w}$ and $x^b \in P'^{t \times h \times w}$ represent the two-layer feature map that is to be fused. $y \in P''^{t \times h \times w}$ represents the merged feature map, t represents the number of frames, and h, w represent the height and width, respectively, of the corresponding feature map.

Concatenation. Cascading the feature maps of two 3D convolutional layers. This can be represented as Equation (4):

$$\begin{cases} y^{cat}_{i,j,2k} = x^{a}_{i,j,k} \\ y^{cat}_{i,j,2k+1} = x^{b}_{i,j,k} \end{cases} 1 \leq i \leq h; 1 \leq j \leq w; 1 \leq k \leq t, \quad (4)$$

where (i, j, k) represents a coordinate point in the feature map.

Sum. Adding the elements of the same coordinate point in the two feature maps. This can be represented as Equation (5):

$$y^{\text{sum}}_{i,j,k} = x^{\text{a}}_{i,j,k} + x^{\text{b}}_{i,j,k} \quad (5)$$

Maximum. Take the larger of the same coordinate points in the two feature maps. This can be represented as Equation (6):

$$y^{\text{max}}_{i,j,k} = \max\{x^{\text{a}}_{i,j,k}, x^{\text{b}}_{i,j,k}\} \quad (6)$$

Average. Calculate the mean of the same coordinate points in the two feature maps. This can be represented as Equation (7):

$$y^{\text{ave}}_{i,j,k} = (x^{\text{a}}_{i,j,k} + x^{\text{b}}_{i,j,k})/2 \quad (7)$$

It is worth noting that the three methods of sum, maximum and average do not change the number of channels. Although the cascade fusion increases the number of channels of the feature map, the information volume of each channel is not compressed. Therefore, the cascading fusion is more conducive to retaining the original information.

4. Building Dataset

4.1. Data Collection

Due to the particularity of battlefield target aggregation behavior, intelligence videos are difficult to obtain. Our article does not consider the disturbances of complex environmental factors (such as complex weather, etc.). Therefore, we collected video data on a satellite simulation platform and built the dataset based on the video data. Behavioral simulation data needs to ensure a visual similarity and behavioral similarity as much as possible. In terms of visual similarity, based on the acquisition of the static images in the open network, Figure 4 shows that the simulated data set has a visual similarity with the real data. In terms of behavioral similarity, the aggregate behavior can be seen as the behavior of groups moving to a certain area. From the perspective of space, the aggregation behavior shows a relative position change in the battlefield space; from the perspective of time, the aggregation behavior is a time series of the density of the combat unit from low to high. The warships are not affected by the terrain, so the trajectory of each target in the gathering behavior is roughly a straight line. Therefore, the data of our simulation has a certain authenticity.

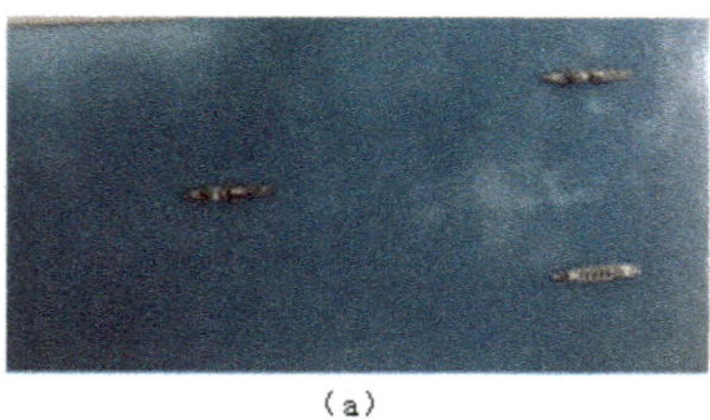
(a)

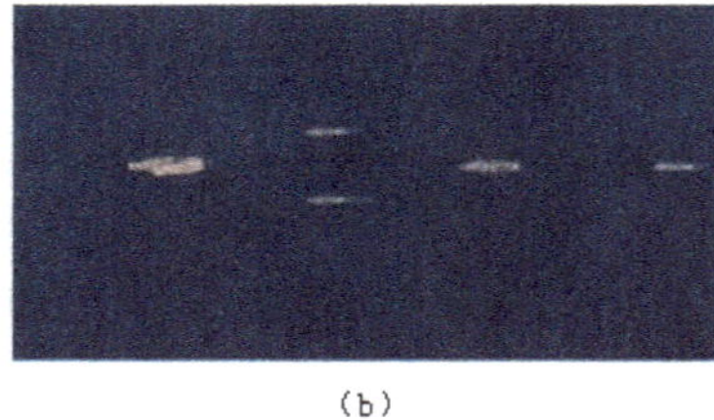
(b)

Figure 4. (**a**) Simulation data. (**b**) Real data. Visually speaking, the background and the shape of the object are similar.

The purpose of this paper is to solve the binary classification problem of behavior. The dataset is divided into two categories: aggregation and others. The aggregate classes include behaviors such as dual-target aggregation and multi-target aggregation. Other classes include actions such as target static and target travel. The aggregate behavior class is the positive samples of model training. Other behavioral classes are negative samples of model training. In reality, in addition to the aggregation behavior, most of the behaviors captured in the video are target static, no target, and target travel (no aggregation trend), so the dataset constructed in this paper is representative. Part of the RGB training samples is shown in Figure 5.

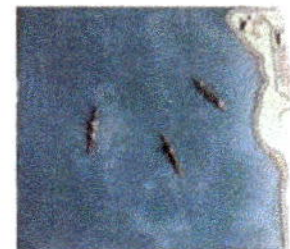
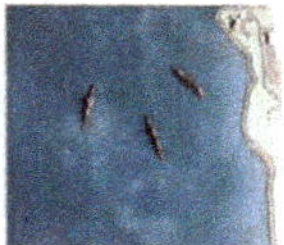
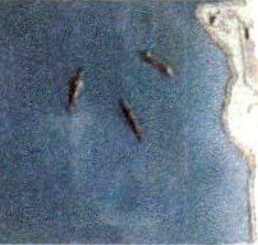

Figure 5. Dataset RGB sample. The above pictures show a few frames extracted from a video clip.

According to the actual speed of 20 times, 1000 segments of the video are collected. Each video lasts for 30 to 40 seconds. The resolution of each video is 720 pixels × 480 pixels, and the frame rate is 25 fps. Our dataset consists of 500 aggregation behavior video and 500 other behavior videos. Each category is randomly divided into five splits, each containing 100 videos. The data set is shown in Table 2.

Table 2. Experimental dataset. The dataset consists of 500 aggregation behavior videos and 500 other behavior videos. Each category is randomly divided into five splits, each containing 100 videos.

Category	Split1	Split2	Split3	Split4	Split5
Aggregation	100	100	100	100	100
Others	100	100	100	100	100

4.2. Optical Flow Video

The optical flow can characterize the motion of the object. Using the optical flow data as the training data is beneficial to improving the recognition rate. On the one hand, the optical flow data can shield the interference of the complex background; on the other hand, the optical flow graph contains enough motion information, which can make the network more comprehensively learn aggregation behavior characteristics. The partial optical flow samples are shown in Figure 6.

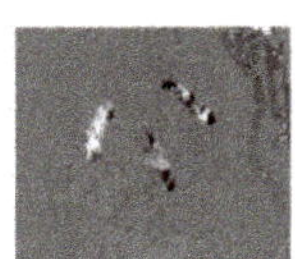
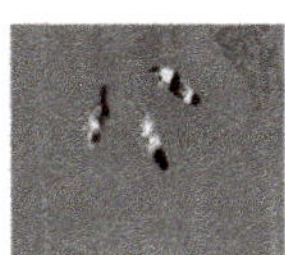
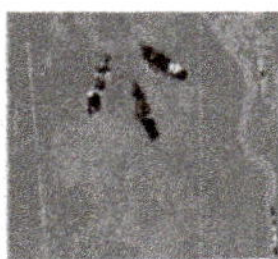

Figure 6. Dataset optical flow sample. The optical flow image is obtained by calculating the optical flow between two frames of the RGB images.

4.3. Data Enhancement

When training deep networks, it is easy to over-fit due to insufficient labeling samples [20]. Data enhancement can effectively avoid over-fitting. This article uses two enhancement strategies for data. (1) In the video, we extract multiple times with different frames as the first frame, and the sampling interval is 30 frames. The extracted segments have 16 frames, and there is an overlap between the extracted segments; (2) Corner cropping. First, we scale the image size to 256 pixels × 256 pixels, and then crop from the center and 4 diagonal regions into 5 sub-images of 224 pixels × 224 pixels.

The experimental results show that the data enhancement improves the generalization ability of the network and improves the recognition accuracy.

5. Experimental Section and Analysis

5.1. Experimental Setup and Training Strategy

We tested our methods in the Ubuntu16.04 environment. The details are shown in Table 3.

Table 3. Experimental configuration.

Operating System	Ubuntu16.04
CPU	Intel Core I9-7940X
GPU	Nvidia GeForce TITANV
Design language	Python3.5
Frame	Pytorch
CUDA	Cuda8.0

The training optimization of the RGB network and optical flow network is based on Back-propagation (BP). Our models are optimized with a vanilla synchronous SGD algorithm with the momentum of 0.9. The networks are trained with an initial learning rate of 0.1, which decays step-wisely with a factor of 0.1. The batch size of the dataset is 16, which is to say that the network has 16 segments per iteration, and the network reaches a steady state when iterating 6000 times. The optical flow is calculated by TVL1 of the OpenCV algorithm.

5.2. Multi-Scale Network Comparison Test

As shown in Figure 7, we compare the accuracy rates in different fusion methods. The fusion methods include Concatenation, Sum, Max, and Ave. The experimental results show that the fusion of concatenation has the highest recognition accuracy.

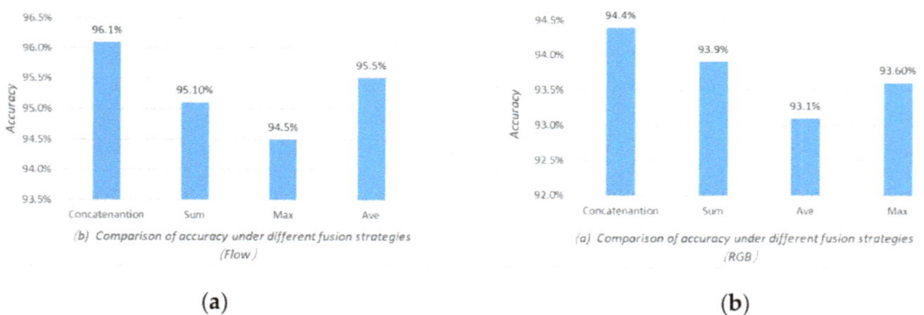

(a) (b)

Figure 7. (**a**) Accuracy of optical flow branches under different fusion strategies. (**b**) Accuracy of RGB branches under different fusion strategies. We compared the accuracy rates in different fusion methods. The experimental result is the average of the results of 5-fold cross-validation tests. Concatenation works best because the concatenation method does not merge the channels and can retain the original information of the feature map.

The two-stream convolutional network in this paper is divided into the RGB network and optical flow network. The network input is 16 frames of the RGB sequence or 16 frames of the optical stream sequence. In order to verify the advantages of this network framework in identifying accuracy and computational complexity, we tested the identification results of the networks in the dataset constructed in this paper. We use FLOPs (floating-point multiplication-adds) to measure the amount of computation.

The experimental comparison method is the I3D [14] network (including the RGB tributary and optical stream tributaries) and MF-Net [18]. The experimental results are shown in Table 4 and Figure 8. The M3D proposed in this paper achieves a 94.4% and 96.1% accuracy, respectively, when the input is the RGB sequence or the optical flow sequence. Comparing the three networks, the recognition results of the proposed algorithm are improved to different degrees than I3D and MF-Net. The experimental results show that: (1) Introducing multi-scale feature fusion into the 3D-CNN can effectively improve the accuracy of the network recognition. (2) The use of multi-fiber modules can greatly reduce the parameter amount and calculation expenses while ensuring the recognition accuracy.

Table 4. Branch network comparison on the dataset constructed in this paper. Test1–5 represents the results of the first to fifth cross-validation tests.

Method	Test1	Test2	Test3	Test4	Test5	Average
RGB I3D	94.5%	90.5%	91.5%	92.5%	93%	92.4%
Flow I3D	95%	92%	93%	94.5%	95%	93.9%
RGB MF-Net	90%	89.5%	88.5%	91%	89%	89.6%
Flow MF-Net	91.5%	90.5%	90.5%	92.5%	90%	91%
RGB M3D (Ours)	96%	92.5%	94%	95%	94.5%	94.4%
Flow M3D (Ours)	97%	94%	95.5%	97.5%	96.5%	96.1%

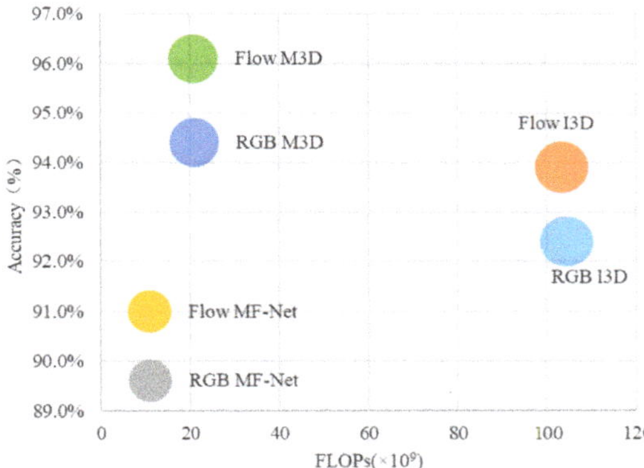

Figure 8. Efficiency comparison between different 3D convolutional networks on the dataset constructed in this paper. The computational complexity is measured using FLOPs, i.e., floating-point multiplication-adds. The area of each circle is proportional to the total parameter number of the model. FLOPs for computing the optical flow are not considered.

5.3. Test of Two-Stream Network

In this section, the experimental results obtained by the M3D(RGB) and the M3D(Flow) are weighted and fused according to different weights, and then the optimal distribution ratio is selected. The results are shown in Table 5, where the distribution ratio = M3D(RGB): M3D(Flow).

Table 5. The accuracy of the Two-Stream network at different weight ratios, where the distribution ratio = M3D(RGB): M3D(Flow). Test1–5 represents the results of the first to fifth cross-validation tests.

Distribution Ratio	Test1	Test2	Test3	Test4	Test5	Average
3:7	97%	94.5%	95.5%	97.5%	96.5%	96.2%
4:6	97%	95.5%	96%	98%	97%	96.7%
5:5	98%	96.5%	96%	98.5%	97.5%	97.3%
6:4	96.5%	94%	95.5%	97%	96.5%	95.9%

As shown in Table 5, when the distribution ratio is 5:5, the Two-Stream network of this paper achieves the best recognition rate. The recognition accuracy on the dataset constructed in this paper reached 97.3%.

5.4. Comparison of Various Methods

Table 6 shows the video action recognition results of different models trained on the dataset established in this paper. It can be seen that: (1) From the perspective of the network structure, the algorithm combines different size feature maps, which is not available in other networks. The fusion of a multi-scale feature map compensates for the risk of the video information loss; (2) From the experimental results, compared with the traditional single-channel input network such as C3D [14], the accuracy of the two-stream M3D is improved by 13.3%. The main reason for this is that the multi-scale feature fusion strategy can more effectively extract spatio-temporal information. Compared with the LSTM+CNN [14], the accuracy of the Two-Stream M3D is improved by 17%. The main reason is that the LSTM (Long Short-Term Memory) on features from the last layers of ConvNets can model a high-level variation, but may not be able to capture fine low-level motion. Because CNN loses many fine-grained underlying features during down-sampling, these underlying features may be critical for a proper motion recognition. That is to say, the feature map outputted by CNN loses the information of smaller targets. If this feature map is input into LSTM for a timing analysis, the recognition result will be greatly affected. Compared with the two-stream network, the Two-Stream M3D proposed in this paper effectively improves the recognition accuracy, which is 1.6% and 5.1% higher than Two-Stream I3D and Two-Stream MF-Net, respectively. Figure 9 shows that the proposed model achieves a good trade-off between computational complexity and accuracy.

Table 6. Action recognition accuracy on the dataset constructed in this paper. Test1–5 represents the results of the first to fifth cross-validation tests. In the input, R means the input is RGB, R+OF means the input is RGB and Optical Flow. The Two-Stream M3D outperforms C3D by 13.3%, LSTM+CNN by 17%, Two-Stream I3D by 1.6%, and Two-Stream MF-Net by 5.1%.

Method	Inputs	Params	Test1	Test2	Test3	Test4	Test5	Average
LSTM+CNN	R	9M	82.5%	81%	77%	82%	79%	80.3%
C3D	R	79M	83%	84%	82.5%	85%	85.5%	84%
Two-Stream I3D	R+OF	23.4M	96%	94%	95%	97%	96.5%	95.7%
Two-Stream MF-Net	R+OF	16.9M	92.5%	92%	91.5%	94%	91%	92.2%
Two-Stream M3D	R+OF	20.9M	98%	96.5%	96%	98.5%	97.5%	97.3%

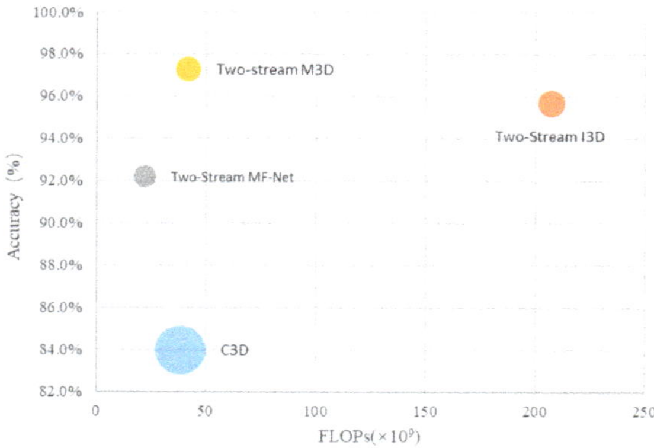

Figure 9. Efficiency comparison between different methods. The computational complexity is measured using FLOPs, i.e., floating-point multiplication-adds. The area of each circle is proportional to the total parameter number of the model. FLOPs for the computing optical flow are not considered.

6. Conclusions

This paper was aimed at the problem of battlefield target aggregation behavior recognition based on intelligence videos: (1) This paper proposes the M3D model, which effectively reduces the impact of network down-sampling on the network recognition accuracy by combining different scale feature maps. The algorithm can effectively deal with the problem that the target of aggregation behavior is small and that the duration is uncertain; (2) Thanks to the multi-fiber module, our algorithm achieves a good trade-off between computational complexity and accuracy. On the established aggregation behavior dataset, the algorithm of this paper is experimentally verified and compared with several advanced algorithms. The results of multiple experiments show that the proposed algorithm can effectively improve the accuracy of the aggregation behavior recognition.

We have not yet verified the algorithm in a complex environment. Under realistic conditions, complex environmental factors will increase the difficulty of recognition of the algorithm. When the cloud is occluded or the visibility is low, the recognition fails because the target cannot be observed. For the ocean target, when the sea surface has a severe diffuse reflection, the optical flow images extracted by the traditional optical flow method will have a large noise, which is not conducive to the algorithm's recognition of the behavior. In future work, in view of the feature extraction advantages of traditional algorithms in complex environments, we will start with a combination of traditional methods and deep learning to explore a more robust recognition algorithm.

Author Contributions: Methodology, software, and validation, H.J.; data curation, formal analysis, Y.P.; writing, review and editing, J.Z.; supervision, H.Y.

Funding: This research received no external funding.

Conflicts of Interest: All authors declare no conflict of interest.

References

1. Huang, M.; Su, S.Z.; Cai, G.R.; Zhang, H.B.; Cao, D.; Li, S.Z. Meta-action descriptor for action recognition in RGBD video. *IET Comput. Vis.* **2017**, *11*, 301–308. [CrossRef]
2. Peng, X.; Wang, L.; Wang, X.; Qiao, Y. Bag of visual words and fusion methods for action recognition: Comprehensive study and good practice. *Comput. Vis. Image Underst.* **2016**, *150*, 109–125. [CrossRef]
3. Duta, I.C.; Ionescu, B.; Aizawa, K.; Sebe, N. Spatio-Temporal VLAD Encoding for Human Action Recognition in Videos. In Proceedings of the International Conference on the MultiMedia Modeling (MMM), Reykjavik, Iceland, 4–6 January 2017.
4. Li, C.; Zhong, Q.; Xie, D.; Pu, S. Collaborative Spatio-temporal Feature Learning for Video Action Recognition. Available online: https://arxiv.org/pdf/1903.01197 (accessed on 14 May 2019).
5. Zolfaghari, M.; Oliveira, G.L.; Sedaghat, N.; Brox, T. Chained Multi-stream Networks Exploiting Pose, Motion, and Appearance for Action Classification and Detection. In Proceedings of the IEEE International Conference on Computer Vision (ICCV), Venice, Italy, 22–29 October 2017.
6. Wang, H.; Schmid, C. Action Recognition with Improved Trajectories. In Proceedings of the IEEE International Conference on Computer Vision (ICCV), Sydney, Australia, 8–12 April 2013.
7. Simonyan, K.; Zisserman, A. Two-Stream Convolutional Networks for Action Recognition in Videos. *Adv. Neural Inf. Process. Syst.* **2014**, *2*, 111–119.
8. Wang, L.; Xiong, Y.; Wang, Z.; Qiao, Y.; Lin, D.; Tang, X.; Van Gool, L. Temporal Segment Networks: Towards Good Practices for Deep Action Recognition. In Proceedings of the European Conference on Computer Vision (ECCV), Amsterdam, The Netherlands, 8–16 October 2016.
9. Feichtenhofer, C.; Pinz, A.; Wildes, R.P. Spatiotemporal Residual Networks for Video Action Recognition. *Neural Inf. Process. Syst.* **2016**, *29*, 3468–3476.
10. Feichtenhofer, C.; Pinz, A.; Wildes, R.P. Spatiotemporal Multiplier Networks for Video Action Recognition. In Proceedings of the IEEE Conference on Computer Vision and Pattern Recognition (CVPR), Honolulu, HI, USA, 22–25 July 2017.
11. Zhu, Y.; Lan, Z.; Newsam, S.; Hauptmann, A.G. Hidden Two-Stream Convolutional Networks for Action Recognition. Available online: https://arxiv.org/pdf/1704.0389 (accessed on 14 May 2019).

12. Donahue, J.; Anne Hendricks, L.; Guadarrama, S.; Rohrbach, M.; Venugopalan, S.; Saenko, K.; Darrell, T. Long-Term Recurrent Convolutional Networks for Visual Recognition and Description. *TPAMI* **2017**, *39*, 677–691. [CrossRef] [PubMed]
13. Varol, G.; Laptev, I.; Schmid, C. Long-Term Temporal Convolutions for Action Recognition. *IEEE Trans. Pattern Anal. Mach. Intell.* **2018**, *40*, 1510–1517. [CrossRef] [PubMed]
14. Carreira, J.; Zisserman, A. Quo Vadis, Action Recognition? In Proceedings of the IEEE Conference on Computer Vision and Pattern Recognition (CVPR), Honolulu, HI, USA, 22–25 July 2017.
15. Qiu, Z.; Yao, T.; Mei, T. Learning Spatio-Temporal Representation with Pseudo-3D Residual Networks. In Proceedings of the IEEE International Conference on Computer Vision (ICCV), Venice, Italy, 22–29 October 2017.
16. Xie, S.; Sun, C.; Huang, J.; Tu, Z.; Murphy, K. Rethinking Spatiotemporal Feature Learning: Speed-Accuracy Trade-offs in Video Classification. In Proceedings of the European Conference on Computer Vision (ECCV), Munich, Germany, 8–14 September 2018; pp. 305–321.
17. Tran, D.; Wang, H.; Torresani, L.; Ray, J.; LeCun, Y.; Paluri, M. A Closer Look at Spatiotemporal Convolutions for Action Recognition. In Proceedings of the IEEE Conference on Computer Vision and Pattern Recognition (CVPR), Salt Lake City, UT, USA, 19–21 June 2018.
18. Chen, Y.; Kalantidis, Y.; Li, J.; Yan, S.; Feng, J. Multi-Fiber Networks for Video Recognition. In Proceedings of the European Conference on Computer Vision (ECCV), Munich, Germany, 8–14 September 2018.
19. Lin, T.Y.; Dollar, P.; Girshick, R.; He, K.M.; Hariharan, B.; Belongie, S. Feature Pyramid Networks for Object Detection. Available online: https://arxiv.org/pdf/1612.03144 (accessed on 14 May 2019).
20. Qu, L.; Wang, K.R.; Chen, L.L.; Li, M.J. Fast road detection based on RGBD images and convolutional neural network Acta Optica Sinica. *Acta Opt. Sinica* **2017**, *37*, 1010003.

Article

A Robust Color Image Watermarking Algorithm Based on APDCBT and SSVD

Xiaoyan Yu, Chengyou Wang * and Xiao Zhou

School of Mechanical, Electrical and Information Engineering, Shandong University, Weihai 264209, China; xyy@mail.sdu.edu.cn (X.Y.); zhouxiao@sdu.edu.cn (X.Z.)

* Correspondence: wangchengyou@sdu.edu.cn; Tel.: +86-631-568-8338

Received: 1 August 2019; Accepted: 11 September 2019; Published: 2 October 2019

Abstract: With the wide application of color images, watermarking for the copyright protection of color images has become a research hotspot. In this paper, a robust color image watermarking algorithm based on all phase discrete cosine biorthogonal transform (APDCBT) and shuffled singular value decomposition (SSVD) is proposed. The host image is transformed by the 8×8 APDCBT to obtain the direct current (DC) coefficient matrix, and then, the singular value decomposition (SVD) is performed on the DC matrix to embed the watermark. The SSVD and Fibonacci transform are mainly used at the watermark preprocessing stage to improve the security and robustness of the algorithm. The watermarks are color images, and a color quick response (QR) code with error correction mechanism is introduced to be a watermark to further improve the robustness. The watermark embedding and extraction processes are symmetrical. The experimental results show that the algorithm can effectively resist common image processing attacks, such as JPEG compression, Gaussian noise, salt and pepper noise, average filter, median filter, Gaussian filter, sharpening, scaling attacks, and a certain degree of rotation attacks. Compared with the color image watermarking algorithms considered in this paper, the proposed algorithm has better performance in robustness and imperceptibility.

Keywords: color image watermarking; copyright protection; robustness; all phase discrete cosine biorthogonal transform (APDCBT); shuffled singular value decomposition (SSVD); Fibonacci transform

---

1. Introduction

With the rapid development of the information age, digital images, as important information carriers, have been widely spread on the network. Images are easy to copy and spread in large quantities, which brings convenience to people, but also makes piracy and infringement more and more serious. In recent years, endless piracy and infringement incidents have emerged, seriously affecting the legitimate rights and interests of copyright owners and disrupting the normal market order and social public order. Improving the level of copyright protection and information security for digital images has become a research hotspot. Techniques for copyright protection include cryptography [1], steganography [2], digital fingerprinting [3], and digital watermarking [4]. Among them, digital watermarking is the most commonly used copyright protection technology for digital products due to its good concealment, robustness, and stability. Until now, digital watermarking has been applied to the copyright protection of texts [5], images [6–8], audios [9], videos [10–12], and so on, especially images.

Since the advent of color displays, black-and-white displays have been rapidly replaced. With the rise of this technology, color images have replaced black-and-white images and gray images. As a result, the watermarking technology for copyright protection of color images has gradually developed. Depending on the difference of the embedding locations, image watermarking can be divided into two categories: watermarking in the spatial domain [13–15] and watermarking in the transform domain [16–28]. Chou and Wu [13] embedded the color watermark image into the color host images uniformly by modifying the quantization indices of color pixels. This algorithm cannot effectively resist various attacks. Su et al. [14] calculated the approximate maximum eigenvalue of the Schur decomposition in the spatial domain to embed the watermark without true Schur decomposition. Su et al. [15] embedded the watermark into the direct current (DC) coefficient of two-dimensional (2D) discrete Fourier transform (DFT) obtained in the spatial domain. These two algorithms have both the advantages of the spatial domain and frequency domain.

Watermarking algorithms in the spatial domain are easy to implement but have poor robustness, while watermarking algorithms in the transform domain are robust and have wide application range. Barni et al. [16] embedded the watermark by changing all discrete cosine transform (DCT) coefficients of each color channel. A threshold was used in this algorithm to minimize the difference of extracted watermark information. Liu [17] extended the perceptual model used in gray images to the color images, and estimated the noise attack threshold of each wavelet coefficient in the luminance and chrominance of the color image to meet the requirements of imperceptibility and robustness. Patvardhan et al. [18] took the quick response (QR) code as the carrier of watermark information, and combined discrete wavelet transform (DWT) and singular value decomposition (SVD) to embed the watermark into the luminance component of host images. Roy and Pal [19] proposed a color multiple watermarking method based on DCT and repetition code. Two watermarks were embedded into green and blue components, respectively. The computational complexity of this algorithm is high. Based on the local invariant significant bit-plane histogram, Niu et al. [20] presented a new robust color image watermarking algorithm, which embedded the watermark into the affine invariant local feature regions. Its limitation is that it has lower watermark capacity. In addition, it cannot be used effectively in real-time applications. Liu et al. [21] proposed a robust hybrid color image watermarking algorithm by fusing a local KAZE feature-based watermarking scheme with a conventional watermarking scheme based on integer wavelet transform (IWT). Its watermark capacity is limited. Chang et al. [22] converted the color host image into the YIQ color space, and embedded two binary watermarks into the DCT domain of the quadrature chrominance component. The imperceptibility of this algorithm is relatively poor. Lakrissi et al. [23] proposed a novel dynamic color image watermarking algorithm combining the DWT with SVD. The watermark was randomly embedded into LL sub-bands based on the human visual system (HVS) to choose an adaptive scaling factor. Li et al. [24] proposed a novel color image watermarking scheme based on the quaternion Hadamard transform (QHT) and Schur decomposition. Rosales-Roldan et al. [25] used SVD, DWT, and DCT to transform the luminance component of host images and embedded the watermark using the quantization index modulation (QIM). Roy and Pal [26] embedded the watermark into the luminance component of host images based on DWT and SVD. Vaidya and Mouli [27] used multiple decompositions, including DWT, contourlet transform (CT), Schur decomposition, and SVD, to embed the watermark into the luminance component of the host image. Mohammad and Gholamhossein [28] adopted the teaching-learning-based optimization (TLBO) method to automatically determine the embedding parameters and suitable locations for inserting the watermark. Laur et al. [29] proposed a non-blind color image watermarking scheme based on DWT, chirp Z-transform (CZT), QR decomposition, SVD, and entropy, which embedded the watermark into low entropy parts of all three RGB components. Cedillo-Hernandez et al. [30] proposed a robust-encoded color image watermarking algorithm using the image normalization procedure and DCT transform. To improve the security of the algorithm, a convolutional encoder was used to encode the watermark. All of these algorithms used gray or binary images as the watermark.

With the development of color image watermarking algorithms, color images have gradually been implemented as watermarks [31–33]. Su et al. [31] performed two-level DCT on the color host image and embedded the color watermark into DC coefficients and alternating current (AC) coefficients. Su and Chen [32] proposed an improved dual color image watermarking scheme based on the Schur decomposition, using the features obtained by the Schur decomposition to embed or extract the watermark images. Pandey et al. [33] embedded the singular values of the watermark into the luminance component of host images using variable scaling factors. This algorithm has high payload capacity, while it is also a non-blind watermarking algorithm. Jia et al. [34] proposed a color image watermarking algorithm based on DWT and QR decomposition, using the features obtained by QR decomposition to embed the watermark. Taking color images as watermarks poses more challenges to the watermark capacity and the imperceptibility of the algorithm.

In this paper, a robust color image watermarking algorithm based on all phase discrete cosine biorthogonal transform (APDCBT) and shuffled singular value decomposition (SSVD) is proposed. The better low-frequency energy accumulation characteristics of the APDCBT and the stability of SSVD are combined to ensure the robustness of the algorithm. Firstly, the host image is divided into nonoverlapping 8 × 8 blocks, and the APDCBT is performed to obtain the DC coefficient matrix. Then, SVD is performed on the DC matrix to embed the watermark. The SSVD is mainly used at the watermark preprocessing stage to improve the security and robustness of the algorithm.

The main contributions of the proposed algorithm are as follows. (1) The APDCBT transform is introduced into the color image watermarking algorithm, and its better low-frequency energy accumulation characteristic is used to find the watermark embedding locations. (2) Fibonacci transform with larger key space is used to improve the security of the algorithm. (3) The watermarks for embedding are color images, not gray or binary images. Additionally, the color QR code with error correction mechanism is introduced to be a watermark, which can further improve the robustness of the algorithm. (4) The proposed algorithm has high robustness to common image processing attacks, such as JPEG compression, Gaussian noise, salt and pepper noise, average filter, median filter, Gaussian filter, sharpening, and scaling attacks.

The rest of this paper is organized as follows. Preliminary concepts are provided in Section 2, including the concepts and characteristics of APDCBT, SSVD, Fibonacci transform, and QR code. Section 3 details the proposed algorithm, including the preprocessing, embedding, and extraction of the watermark. The experimental results and discussions are presented in Section 4. Finally, the conclusions and future work are discussed in Section 5.

2. Preliminary Concepts

To improve the robustness of the algorithm, the advantages of the APDCBT and SSVD are combined to determine the appropriate locations for watermark embedding. To improve the security of the algorithm, the Fibonacci transform is adopted to scramble the watermark image. In addition, the QR code is introduced to further improve the robustness of the algorithm. This section mainly introduces the concepts and characteristics of these three transforms and the QR code.

2.1. APDCBT

The APDCBT was proposed to solve the serious block artifacts caused by DCT at low bit rates in image compression [35]. Compared with DCT, the APDCBT has better low-frequency energy accumulation and high-frequency energy attenuation characteristics. Due to these characteristics, the APDCBT has been applied in many fields, such as image compression and digital watermarking [36]. The transform matrix of APDCBT can be expressed as Equation (1):

$$B(i,j) = \begin{cases} \frac{N-i}{N^2}, & i = 0,1,\cdots,N-1,\ j = 0 \\ \frac{1}{N^2}\left[(N-i)\cos\frac{ij\pi}{N} - \csc\frac{j\pi}{N}\sin\frac{ij\pi}{N}\right], & i = 0,1,\cdots,N-1,\ j = 1,2,\cdots,N-1 \end{cases}, \tag{1}$$

where N is the size of the image block $\boldsymbol{X}$. Performing the APDCBT on $\boldsymbol{X}$ can be presented as Equation (2):

$$\boldsymbol{A} = \boldsymbol{B}\boldsymbol{X}\boldsymbol{B}^{\mathrm{T}}, \tag{2}$$

where $\boldsymbol{A}$ is the transform coefficient matrix and $\boldsymbol{B}$ is the transform matrix of APDCBT.

2.2. SSVD

The SSVD [37] was developed on the basis of the standard SVD. Different from the SVD, SSVD adds a scrambling process, which can be expressed as $\boldsymbol{Y} = \hat{S}\{\boldsymbol{X}\}$, where $\boldsymbol{X}$ is the input image, $\boldsymbol{Y}$ is the scrambled image, and $\hat{S}$ is the shuffled operator. In other words, SSVD scrambles the input image first, and then performs standard SVD on the scrambled image $\boldsymbol{Y}$. The scrambling method used in the proposed algorithm is the Fibonacci transform, which will be introduced in Section 2.3.

SVD is a type of matrix decomposition that can be applied to any matrix. By performing SVD on a complex matrix, three smaller and simpler submatrices can be obtained, which can describe the important characteristics of the complex matrix. For a matrix $\boldsymbol{Y}$, its SVD can be expressed as Equation (3):

$$\boldsymbol{Y} = \boldsymbol{U}\boldsymbol{S}\boldsymbol{V}^{\mathrm{T}}, \tag{3}$$

where $\boldsymbol{U}$ and $\boldsymbol{V}$ are orthogonal matrices and $\boldsymbol{S}$ is a diagonal matrix. The matrix $\boldsymbol{S}$ contains the singular values of matrix $\boldsymbol{Y}$, which can represent the main information of matrix $\boldsymbol{Y}$ and has high stability. Due to this characteristic, SVD is widely used in the watermarking field.

2.3. Fibonacci Transform

Matrix transform is one of the most commonly used image scrambling methods. For an image $\boldsymbol{F}$ of size $N \times N$, its Fibonacci transform [11] can be shown as Equation (4):

$$\begin{bmatrix} x' \\ y' \end{bmatrix} = \begin{bmatrix} 1 & 1 \\ 1 & 0 \end{bmatrix} \begin{bmatrix} x \\ y \end{bmatrix} \mathrm{mod}(N), \tag{4}$$

where (x, y) is the coordinate of image $\boldsymbol{F}$ and (x', y') is the coordinate after the transform. The Fibonacci transform is periodic, and its transform periods for images with different sizes are shown in Table 1. When the images are the same size, the period of Fibonacci transform is twice as long as an Arnold transform [38]. Therefore, the Fibonacci transform can provide more key space and improve the security of watermarking algorithms.

Table 1. The periods of the Fibonacci transform for images with different sizes.

Size of Image	4×4	8×8	16×16	32×32	64×64	128×128	256×256
Period	6	12	24	48	96	192	384

2.4. QR Code

The QR code is a two-dimensional bar code that can be read quickly. Compared with the traditional bar code, the QR code can store more data information. Due to this characteristic, QR codes have been widely used in the information identification of intelligent devices. The size of QR codes can range from 21×21 to 177×177, mainly including data area, alignment pattern, timing pattern, finder pattern, quit zone, dark module, and separator. Since QR codes have four levels of error correction capability and its maximum fault tolerance can reach 30%, it can still be read correctly even if it is damaged. In addition, QR codes also have strong bending resistance.

Due to the error correction algorithm of the QR code, it can ensure the security of the information to be hidden. In watermarking algorithms, embedding the hidden information into the QR code and using the generated QR code as the watermark can improve the robustness of the algorithm. Even if the extracted QR code image is destroyed, the hidden information can still be decoded.

3. Proposed Color Image Watermarking Algorithm

In this section, the specific steps of the proposed color image watermarking algorithm are explained. The algorithm consists of three parts: watermark preprocessing, watermark embedding, and watermark extraction. The APDCBT is introduced into the color image watermarking algorithm, and its better low-frequency energy accumulation characteristic is used to find the watermark embedding locations, which is the main contribution of the proposed algorithm. In addition, the Fibonacci transform and the QR code are used to improve the security and the robustness of the proposed algorithm, which are also the contributions of the proposed algorithm. The frameworks of watermark embedding and watermark extraction are shown in Figure 1, and the specific steps are explained in the following subsections.

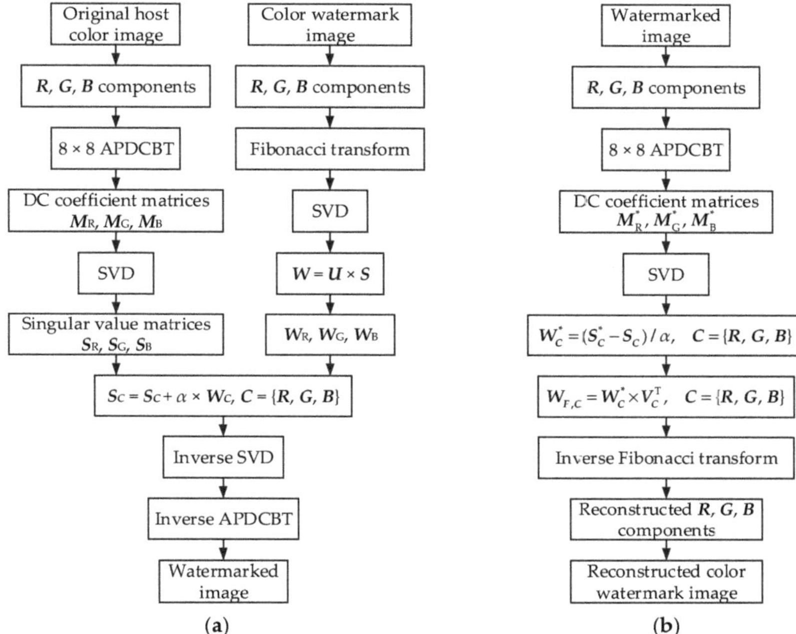

Figure 1. The frameworks of the proposed color image watermarking algorithm: (**a**) Watermark embedding; (**b**) Watermark extraction. APDCBT: all phase discrete cosine biorthogonal transform; SVD: singular value decomposition; DC: direct current.

3.1. Watermark Preprocessing

The watermark preprocessing is performed to improve the security of the watermarking algorithm. SSVD is used to preprocess the watermark image. Here, the Fibonacci transform is used to scramble the color watermark image, and the scrambling process is shown in Figure 2.

The specific watermark preprocessing process is presented below.

Step 1: Divide the color watermark image $\boldsymbol{W}$ of size 64 × 64 into $\boldsymbol{R}_{\mathrm{W}}$, $\boldsymbol{G}_{\mathrm{W}}$, and $\boldsymbol{B}_{\mathrm{W}}$ components.

Step 2: Apply the Fibonacci transform to scramble the three color components.

Step 3: Perform SVD on the scrambled components, as shown in Equation (5):

$$\boldsymbol{C}_{\mathrm{s}} = \boldsymbol{U}_{\mathrm{Cs}}\boldsymbol{S}_{\mathrm{Cs}}\boldsymbol{V}_{\mathrm{Cs}}^{\mathrm{T}},\ \boldsymbol{C} = \{\boldsymbol{R},\ \boldsymbol{G},\ \boldsymbol{B}\}, \tag{5}$$

where $\boldsymbol{C}_{\mathrm{s}}$ are the three scrambled color components, $\boldsymbol{U}_{\mathrm{Cs}}$ and $\boldsymbol{V}_{\mathrm{Cs}}$ are orthogonal matrices for $\boldsymbol{C}_{\mathrm{s}}$, and $\boldsymbol{S}_{\mathrm{Cs}}$ are singular value matrices for $\boldsymbol{C}_{\mathrm{s}}$.

Step 4: Generate embedded watermark using Equation (6):

$$\boldsymbol{W}_{C} = \boldsymbol{U}_{\mathrm{Cs}}\boldsymbol{S}_{\mathrm{Cs}},\ \boldsymbol{C} = \{\boldsymbol{R},\ \boldsymbol{G},\ \boldsymbol{B}\}, \tag{6}$$

where W_C, namely W_R, W_G, and W_B, are the generated watermarks for embedding.

To more comprehensively illustrate the watermark preprocessing process, its pseudocode is shown in Algorithm 1.

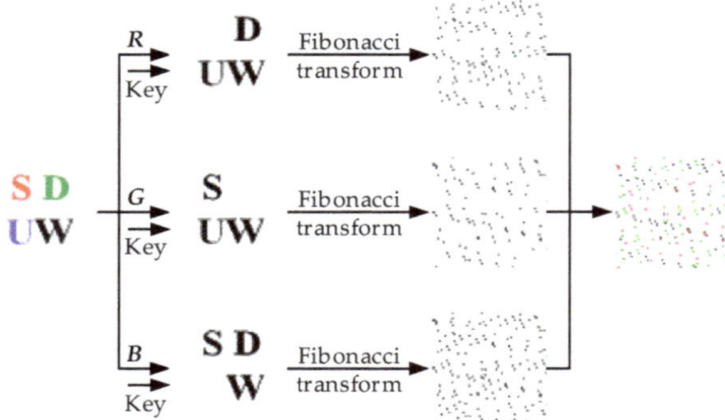

Figure 2. The scrambling process of a color watermark image using the Fibonacci transform.

Algorithm 1. Watermark Preprocessing Algorithm

Variable Declaration:

SDUW: color watermark image

W: read the watermark image

R_W, G_W, and B_W: three components of the color watermark image

R_s, G_s, and B_s: three scrambled components using Fibonacci transform

U_{Rs}, V_{Rs}, U_{Gs}, V_{Gs}, U_{Bs}, and V_{Bs}: orthogonal matrices for R_s, G_s, and B_s

S_{Rs}, S_{Gs}, and S_{Bs}: singular value matrices for R_s, G_s, and B_s

W_R, W_G, and W_B: three watermark components for embedding

Watermark Preprocessing Procedure:

Start Procedure:

1: Read the color watermark image

$W \leftarrow$ SDUW.png (color watermark image of size 64 × 64);

$R_W \leftarrow W$ (:, :, 1); $G_W \leftarrow W$ (:, :, 2); $B_W \leftarrow W$ (:, :, 3);

2: Scramble the three color components of the watermark image

$R_s \leftarrow$ Fibonacci transform (R_W); $G_s \leftarrow$ Fibonacci transform (G_W); $B_s \leftarrow$ Fibonacci transform (B_W);

3: Perform SVD on each scrambled color components

$[U_{Rs}, S_{Rs}, V_{Rs}] \leftarrow$ SVD (R_s); $[U_{Gs}, S_{Gs}, V_{Gs}] \leftarrow$ SVD (G_s); $[U_{Bs}, S_{Bs}, V_{Bs}] \leftarrow$ SVD (B_s);

4: Obtain watermark for embedding

$W_R = U_{Rs} \times S_{Rs}$; $W_G = U_{Gs} \times S_{Gs}$; $W_B = U_{Bs} \times S_{Bs}$;

End Procedure

3.2. Watermark Embedding

The watermark embedding process is shown in Figure 1a, and the detailed steps are given below.

Step 1: Divide the color host image I of size 512 × 512 into R, G, and B components.

Step 2: Divide each component into nonoverlapping 8 × 8 blocks, and apply the APDCBT to each block to obtain the DC coefficient, and then the DC coefficient matrices M_C can be obtained, where $C = \{R, G, B\}$.

Step 3: Perform SVD on M_C to obtain singular value matrices S_C, where $C = \{R, G, B\}$.

Step 4: Use Algorithm 1 to preprocess the color watermark image and obtain W_C, where $C = \{R, G, B\}$.

Step 5: Embed W_C into S_C according to Equation (7):

$$S_{CW} = S_C + \alpha \times W_C,\ S_{CW} = U_{CW} S_{CW1} V_{CW}^{T},\ C = \{R,\ G,\ B\}, \quad (7)$$

where α is the embedding strength of the watermark, and S_{CW} are three singular value matrices with the embedded watermark. U_{CW} and V_{CW} are orthogonal matrices for S_{CW}, and S_{CW1} are the singular value matrices for S_{CW}. Save U_{CW}, V_{CW}, and S_C for watermark extraction process.

Step 6: Apply the inverse SVD to S_{CW1}, then the watermarked M_C^* can be obtained.

Step 7: After performing the inverse APDCBT, three watermarked color components can be obtained. Use the concatenate (cat) operator to stack them, and then the watermarked color image I^* can be obtained.

To more comprehensively illustrate the watermark embedding process, its pseudocode is shown in Algorithm 2.

Algorithm 2. Watermark Embedding Algorithm

Variable Declaration:

Lena: color host image
I: read the color host image
R, G, and B: three color components of the host image
M_C: three DC coefficient matrices obtained by APDCBT, where $C = \{R, G, B\}$
U_C and V_C: orthogonal matrices for M_C
S_C: singular value matrices for M_C
W_C: three preprocessed color components of the watermark image, where $C = \{R, G, B\}$
α: scaling factor
S_{CW}: three singular value matrices with the embedded watermark
U_{CW} and V_{CW}: the orthogonal matrices for S_{CW}
S_{CW1}: the singular value matrices for S_{CW}
M_C^*: watermarked DC coefficient matrices
C^*: three watermarked color components, including R^*, G^*, and B^*
I^*: watermarked image

Watermark Embedding Procedure:

Start Procedure:

1: Read the color host image
$I \leftarrow$ Lena.ppm (color host image of size 512×512);
$R \leftarrow I\,(:, :, 1)$; $G \leftarrow I\,(:, :, 2)$; $B \leftarrow I\,(:, :, 3)$;

2: Perform 8 × 8 APDCBT to three color components
$M_C \leftarrow$ APDCBT (C) //$C = \{R, G, B\}$

3: Perform SVD on M_C
$[U_C, S_C, V_C] \leftarrow$ SVD (M_C)

4: Watermark Embedding
$S_{CW} = S_C + \alpha \times W_C$
$[U_{CW}, S_{CW1}, V_{CW}] \leftarrow$ SVD (M_C)
$M_C^* \leftarrow U_C S_{CW1} V_C^{T}$

5: Perform the inverse APDCBT
$C^* \leftarrow$ inverse APDCBT (M_C^*) //$C^* = \{R^*, G^*, B^*\}$
$I^* \leftarrow$ cat (3, R^*, G^*, B^*)

End Procedure

3.3. Watermark Extraction

The watermark extraction process is shown in Figure 1b, and the detailed steps are given below.

Step 1: Divide the suspicious color image I' into R', G', and B' components.

Step 2: Divide each component into nonoverlapping 8×8 blocks, and apply the APDCBT to each block to obtain the DC coefficient matrices M'_C, where $C = \{R', G', B'\}$.

Step 3: Perform SVD on M'_C to obtain singular value matrices S'_{CW1}, which can be presented as Equation (8):

$$M'_C = U'_C S'_{CW1} \left(V'_C\right)^{\mathrm{T}},\ C = \left\{R', G', B'\right\}, \tag{8}$$

where U'_C and V'_C are orthogonal matrices for M'_C.

Step 4: The embedded watermark W'_C can be obtained according to Equation (9):

$$S'_C = U_{CW} S'_{CW1} (V_{CW})^{\mathrm{T}},\ W'_C = \left(S'_C - S_C\right)/\alpha,\ C = \left\{R', G', B'\right\}, \tag{9}$$

Step 5: Obtain the three color components C'_W using Equation (10):

$$C'_W = W'_C (V_{Cs})^{\mathrm{T}},\ C = \left\{R', G', B'\right\}, \tag{10}$$

Step 6: Apply inverse Fibonacci transform on C'_W to obtain the reconstructed color components C'_{WR}.

Step 7: Reconstruct the final watermark image W' by using the concatenate operator to stack the three color components C'_{WR}, that is, R'_{WR}, G'_{WR}, and B'_{WR}.

To more comprehensively illustrate the watermark extraction process, its pseudocode is shown in Algorithm 3.

Algorithm 3. Watermark Extraction Algorithm

Variable Declaration:

- I': the suspicious color image
- R', G', and B': three color components of I'
- M'_C: three DC coefficient matrices obtained by APDCBT, where $C = \left\{R', G', B'\right\}$
- U'_C and V'_C: orthogonal matrices for M'_C
- S'_{CW1}: singular value matrices for M'_C
- S_C: saved singular value matrices for M_C of host image
- U_{CW} and V_{CW}: saved orthogonal matrices for S_{CW} in watermark embedding process
- S'_C: watermarked singular value matrices
- α: scaling factor
- W'_C: extracted watermark
- C'_W: three reconstructed color components of watermark image
- C'_{WR}: three inverse scrambled color components of watermark image
- W': reconstructed watermark image

Watermark Embedding Procedure:

Start Procedure:

1: **Read the suspicious color image**
   $I' \leftarrow$ Suspicious image.ppm
   $R' \leftarrow I'(:, :, 1)$; $G' \leftarrow I'(:, :, 2)$; $B' \leftarrow I'(:, :, 3)$;

2: **Perform 8 × 8 APDCBT to three color components**
   $M'_C \leftarrow$ APDCBT (C) //$C = \left\{R', G', B'\right\}$

3: **Perform SVD on M'_C**
   $[U'_C, S'_{CW1}, V'_C] \leftarrow$ SVD (M'_C)

4: **Watermark Extraction**
   $S'_C \leftarrow U_{CW} S'_{CW1} (V_{CW})^{\mathrm{T}}$ //$C = \left\{R', G', B'\right\}$
   $W'_C \leftarrow \left(S'_C - S_C\right)/\alpha$
   $C'_W \leftarrow W'_C (V_{Cs})^{\mathrm{T}}$

5: **Apply inverse Fibonacci transform on C'_W**
   $C'_{WR} \leftarrow$ inverse Fibonacci transform (C'_W) //$C = \left\{R', G', B'\right\}$

6: **Reconstruct the final watermark image**
   $W' \leftarrow$ cat $(3, R'_{WR}, G'_{WR}, B'_{WR})$

End Procedure

---

4. Experimental Results and Analysis

In this section, to evaluate the performance of the proposed algorithm, several experiments are performed using MATLAB R2014a on an Intel Core i5-4590 3.30 GHz CPU. All 24-bit color images of size 512×512 in the CVG-UGR image database [39] are prepared as host images. In this paper, six color images, including "Lena", "Baboon", "Airplane", "Sailboat", "House", and "Peppers", are selected to illustrate the experimental results, which are shown in Figure 3. In addition, three 24-bit color images of size 64×64 are used as the watermarks, which are shown in Figure 4. Among them, Figure 4a is the school badge of Shandong University; Figure 4b is the English abbreviation of Shandong University (Weihai) drawn with Microsoft Visio 2010; and Figure 4c is a color QR code containing a personal website [40] generated by a QR code generator. The imperceptibility and robustness of the proposed algorithm are tested and compared with algorithms [18,31,34].

Figure 3. The original host images: (**a**) Lena; (**b**) Baboon; (**c**) Airplane; (**d**) Sailboat; (**e**) House; (**f**) Peppers.

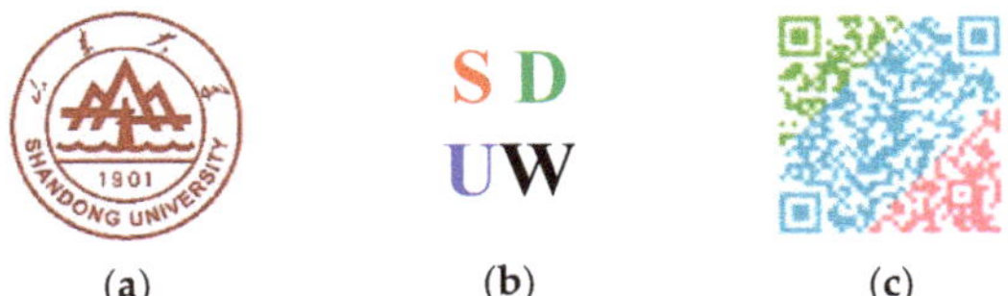

Figure 4. The color watermark images: (**a**) Logo; (**b**) SDUW; (**c**) QR code.

4.1. Evaluation Indexes

In general, peak signal-to-noise ratio (PSNR) and structural similarity index (SSIM) [41] are used to evaluate the imperceptibility of watermarking algorithms, while the normalized correlation coefficient (NCC) is used to evaluate the robustness of watermarking algorithms [31]. The definition of PSNR is shown as Equation (11):

$$\mathrm{PSNR} = 10\log_{10}\frac{255^2}{\mathrm{MSE}}\,(\mathrm{dB}), \tag{11}$$

where the mean square error (MSE) can be defined as Equation (12):

$$\text{MSE} = \frac{1}{M \times N} \sum_{m=1}^{M} \sum_{n=1}^{N} [f(m,n) - f_w(m,n)]^2, \tag{12}$$

where f and f_w are the original host and watermarked images of size $M \times N$, respectively. In this paper, the PSNR of color images is calculated by averaging the PSNRs of R, G, and B components (PSNR_R, PSNR_G, and PSNR_B), which can be defined as Equation (13):

$$\text{PSNR}_c = \frac{1}{3}(\text{PSNR}_R + \text{PSNR}_G + \text{PSNR}_B), \tag{13}$$

where PSNR_c is the PSNR value of color images. The definition of SSIM is shown as Equation (14):

$$\text{SSIM}(f, f_w) = \frac{(2\mu_f \mu_{f_w} + C_1)(2\sigma_{ff_w} + C_2)}{(\mu_f^2 + \mu_{f_w}^2 + C_1)(\sigma_f^2 + \sigma_{f_w}^2 + C_2)}, \tag{14}$$

where f and f_w are the original host and watermarked images, respectively. μ_f and μ_{f_w} represent the mean values of f and f_w, respectively; σ_f and σ_{f_w} are the variances of f and f_w, respectively; σ_{ff_w} is the covariance between f and f_w; and C_1 and C_2 are two constants to maintain the stability. Usually, the larger the PSNR and SSIM are, the better the imperceptibility is.

The definition of NCC is shown as Equation (15):

$$\text{NCC} = \frac{\sum\limits_{p=1}^{3} \sum\limits_{i=1}^{M} \sum\limits_{j=1}^{N} [W(i,j,p) \times W'(i,j,p)]}{\sqrt{\sum\limits_{p=1}^{3} \sum\limits_{i=1}^{M} \sum\limits_{j=1}^{N} [W(i,j,p)]^2} \sqrt{\sum\limits_{p=1}^{3} \sum\limits_{i=1}^{M} \sum\limits_{j=1}^{N} [W'(i,j,p)]^2}}, \tag{15}$$

where W and W' are the original color watermark image and extracted watermark image of size $M \times N$, respectively. The NCC values range from 0 to 1, and the larger the NCC is, the stronger the robustness is.

4.2. Imperceptibility

Imperceptibility means that the watermark embedded into the image cannot be perceived by the human eyes, that is, the embedded watermark cannot affect the visual quality of the image. The embedding strength of the watermark plays an important role in the performance of a watermarking algorithm. The selection of it must take the imperceptibility and robustness of the watermark into account. Here, taking "Lena" as the host image, the changes of PSNR, SSIM, and NCC with the embedding strength α are shown in Figure 5.

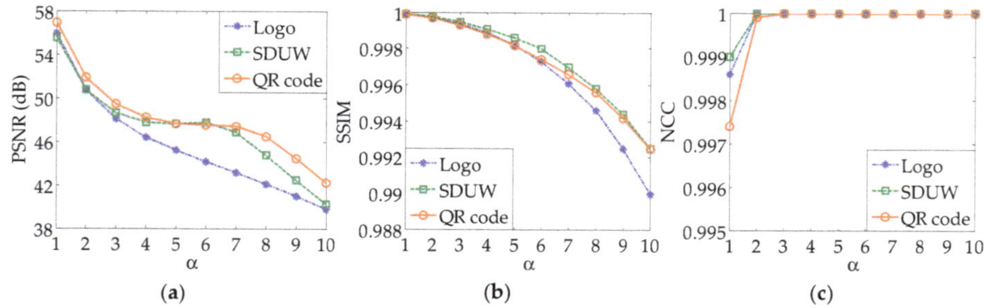

Figure 5. The changes of peak signal-to-noise ratio (PSNR), structural similarity index (SSIM), and normalized correlation coefficient (NCC) with the embedding strength α: (**a**) PSNR; (**b**) SSIM; (**c**) NCC.

To show the imperceptibility of the algorithm more comprehensively, taking "Baboon", "Airplane", "Sailboat", "House", and "Peppers" as objects, the change of the imperceptibility with the embedding strength is tested, and the test results are shown in Table 2.

Table 2. The change of the proposed algorithm's imperceptibility with the embedding strength α.

Host Images	α	Logo			SDUW			QR Code		
		PSNR (dB)	SSIM	NCC	PSNR (dB)	SSIM	NCC	PSNR (dB)	SSIM	NCC
Baboon	1	55.90	1.0000	0.9992	55.49	1.0000	0.9993	56.88	1.0000	0.9980
	3	47.88	0.9998	1.0000	48.25	0.9998	1.0000	49.20	0.9998	1.0000
	5	44.84	0.9992	1.0000	46.53	0.9995	1.0000	46.78	0.9995	1.0000
	7	42.91	0.9982	1.0000	45.97	0.9988	1.0000	45.80	0.9989	1.0000
Airplane	1	55.69	0.9997	0.9979	55.20	0.9998	0.9986	56.63	0.9998	0.9965
	3	47.17	0.9979	1.0000	47.15	0.9984	1.0000	48.32	0.9984	1.0000
	5	43.63	0.9942	1.0000	44.34	0.9955	1.0000	45.15	0.9960	1.0000
	7	41.43	0.9877	1.0000	43.11	0.9914	1.0000	43.40	0.9922	1.0000
Sailboat	1	55.98	0.9999	0.9995	55.46	0.9999	0.9998	56.85	0.9999	0.9992
	3	48.09	0.9993	1.0000	48.18	0.9994	1.0000	49.09	0.9995	1.0000
	5	45.43	0.9979	1.0000	46.52	0.9986	1.0000	46.62	0.9985	1.0000
	7	44.10	0.9960	1.0000	46.56	0.9973	1.0000	45.78	0.9970	1.0000
House	1	55.78	0.9997	0.9991	55.29	0.9998	0.9995	56.71	0.9998	0.9975
	3	47.49	0.9972	1.0000	47.57	0.9984	1.0000	48.60	0.9985	1.0000
	5	44.25	0.9934	1.0000	45.22	0.9963	1.0000	45.73	0.9965	1.0000
	7	42.33	0.9879	1.0000	44.45	0.9932	1.0000	44.36	0.9936	1.0000
Peppers	1	56.13	0.9826	0.9984	55.79	0.9844	0.9990	57.16	0.9843	0.9977
	3	48.80	0.9729	1.0000	49.70	0.9793	1.0000	50.36	0.9775	1.0000
	5	46.34	0.9685	1.0000	48.89	0.9808	1.0000	49.07	0.9777	1.0000
	7	44.13	0.9662	1.0000	46.06	0.9824	1.0000	46.21	0.9806	1.0000

As can be seen from Figure 5 and Table 2, under the same conditions, the imperceptibility of the watermark using a QR code is better than the other two types of watermarks, which shows that QR code can be used as the carrier of watermark information to achieve the watermark embedding. When the embedding strength α equals 5, most PSNR values are larger than 44 dB and most SSIM values are larger than 0.99, while all the NCC values equal 1. Therefore, the embedding strength in the proposed algorithm is set to 5. Taking "Lena" and "Baboon" as objects, the subjective effects of the watermarked images and the extracted watermark images when without performing any attacks on host images are shown in Table 3.

From Table 3, we can see that the watermarked images have no trace of watermark embedding, and the watermarks can be completely extracted. To compare the proposed algorithm with other three robust image watermarking algorithms, the PSNR, SSIM, NCC, and the watermark capacity of them are listed in Table 4.

From Table 4, we can see that the watermarks in algorithm [18] are gray images, while algorithms [31,34] and the proposed algorithm embed color watermark images into color host images. Although the PSNR of the proposed algorithm is less than algorithm [18] and larger than algorithms [31,34], the SSIM and NCC of the proposed algorithm are both better than those of the other three algorithms.

Table 3. The subjective effects of the proposed algorithm with three color watermark images.

Watermarks	Logo	SDUW	QR Code
Watermarked images (PSNR (dB)/SSIM)	45.24/0.9982	47.66/0.9986	47.70/0.9982
Extracted watermarks (NCC)	1.0000	1.0000	1.0000
Watermarked images (PSNR (dB)/SSIM)	44.84/0.9992	46.53/0.9995	46.78/0.9995
Extracted watermarks (NCC)	1.0000	1.0000	1.0000

Table 4. Comparisons of the PSNR, SSIM, NCC, and the watermark capacity among different color image watermarking algorithms.

Algorithms	Patvardhan et al. [18]	Su et al. [31]	Jia et al. [34]	Proposed
Host images	$512 \times 512 \times 3$	$512 \times 512 \times 3$	$512 \times 512 \times 3$	$512 \times 512 \times 3$
Watermark images	gray	color	color	color
PSNR (dB)	52.13	46.07	41.45	49.03
SSIM	0.9975	0.9913	0.9842	0.9992
NCC	0.9962	0.9999	1.0000	1.0000
Capacity	128×128	$64 \times 64 \times 3$	$32 \times 32 \times 3$	$64 \times 64 \times 3$

4.3. Robustness

Robustness means that the watermark image can still be extracted from the watermarked image after suffering various attacks. In the storage and transmission processes, it is inevitable for color images to undergo various attacks. To test the robustness of the proposed algorithm, various common image processing attacks and geometric attacks are performed on the watermarked color images, including JPEG compression, Gaussian noise, salt and pepper noise, average filter, median filter, Gaussian filter, sharpening, rotation, and scaling attacks. Taking "Lena", "Baboon", and "Sailboat" as objects, when "SDUW" is used as the watermark, the results of extracted watermarks and the NCC values under various attacks are shown in Table 5; when "QR code" is used as the watermark, the results of extracted watermarks and the NCC values under various attacks are shown in Table 6.

Table 5. The results of extracted watermarks (SDUW) and the NCC values under various attacks.

Attacks	Parameter	Lena	Baboon	Sailboat
JPEG	QF = 30	S D UW 0.9999	S D UW 0.9999	S D UW 0.9998
	QF = 60	S D UW 1.0000	S D UW 1.0000	S D UW 1.0000
Gaussian noise	0.01	S D UW 0.9997	S D UW 0.9998	S D UW 0.9997
	0.02	S D UW 0.9986	S D UW 0.9994	S D UW 0.9990
Salt and Pepper noise	0.01	S D UW 0.9998	S D UW 0.9999	S D UW 0.9999
	0.02	S D UW 0.9995	S D UW 0.9998	S D UW 0.9997
Average filter	5 × 5	S D UW 0.9991	S D UW 0.9988	S D UW 0.9984
Median filter	5 × 5	S D UW 0.9999	S D UW 0.9988	S D UW 0.9995
Gaussian filter	5 × 5	S D UW 1.0000	S D UW 1.0000	S D UW 1.0000
Sharpening	0.2	S D UW 0.9977	S D UW 0.9971	S D UW 0.9974
Rotation	5°	S D UW 0.9922	S D UW 0.9905	S D UW 0.9941
Scaling	0.5	S D UW 1.0000	S D UW 1.0000	S D UW 1.0000

Table 6. The results of extracted watermarks (QR code) and the NCC values under various attacks.

Attacks	Parameter	Lena	Baboon	Sailboat
JPEG	QF = 30	0.9997	0.9998	0.9996
	QF = 60	0.9999	0.9999	0.9999
Gaussian noise	0.01	0.9991	0.9994	0.9994
	0.02	0.9966	0.9979	0.9978
Salt and Pepper noise	0.01	0.9996	0.9998	0.9999
	0.02	0.9986	0.9993	0.9991
Average filter	5 × 5	0.9978	0.9971	0.9963
Median filter	5 × 5	0.9996	0.9972	0.9988
Gaussian filter	5 × 5	1.0000	0.9999	0.9999
Sharpening	0.2	0.9955	0.9932	0.9941
Rotation	5°	0.9886	0.9855	0.9900
Scaling	0.5	1.0000	1.0000	1.0000

It can be seen from Tables 5 and 6 that the proposed algorithm has high robustness to JPEG compression, noise, filtering, sharpening, and scaling attacks. Although the proposed algorithm can resist a small degree of rotation attacks, it will fail when the rotation angle increases. In the actual image transmission process, images are usually affected by more than one attack. Most color image watermarking algorithms did not consider the robustness when multiple attacks are combined. In this paper, to more comprehensively prove the effectiveness of the proposed algorithm, taking "Lena" as the host image and "QR code" as the watermark, the robustness of the proposed algorithm is compared with algorithms [18,31,34]. In addition, hybrid attacks, which include two or three kinds of attacks, are also considered. The comparison results are shown in Table 7.

Table 7. Comparisons of the NCC values of the watermarks extracted by different robust color image watermarking algorithms.

Attacks	Patvardhan et al. [18]	Su et al. [31]	Jia et al. [34]	Proposed
Gaussian noise (0.02)	0.8104	0.8300	0.8848	0.9966
Salt and Pepper noise (0.02)	0.9335	0.9024	0.9676	0.9986
Average filter (3 × 3)	0.9910	0.8662	0.9485	0.9996
Median filter (3 × 3)	0.9958	0.8697	0.9622	0.9999
Gaussian filter (3 × 3)	0.9960	0.9449	0.9602	1.0000
JPEG (30)	0.9953	0.8437	0.8831	0.9997
Sharpening (0.2)	0.7964	0.8406	0.8943	0.9955
Rotation (5°)	0.9940	0.9666	0.8709	0.9886
Scaling (2)	0.9961	0.9901	0.9997	1.0000
Gaussian noise (0.02) + Average filter (3 × 3)	0.9951	0.8406	0.8724	0.9964
Salt and Pepper noise (0.05) + Median filter (5 × 5)	0.9937	0.8453	0.9202	0.9995
Gaussian noise (0.02) + JPEG (30)	0.8157	0.8394	0.8785	0.9963
Salt and Pepper noise (0.05) + JPEG (50)	0.8402	0.8432	0.8923	0.9934
Average filter (3 × 3) + JPEG (30)	0.9910	0.8413	0.8801	0.9995
Median filter (3 × 3) + JPEG (70)	0.9950	0.8405	0.9062	0.9999
Gaussian noise (0.02) + Scaling (0.5)	0.9907	0.8389	0.8821	0.9967
Salt and Pepper noise (0.05) + Scaling (2)	0.9639	0.8580	0.9668	0.9938
Average filter (5 × 5) + Scaling (0.5)	0.9883	0.8366	0.9002	0.9977
Median filter (5 × 5) + Scaling (2)	0.9917	0.8826	0.9260	0.9996
JPEG (30) + Scaling (0.5)	0.9895	0.8424	0.8822	0.9998
JPEG (50) + Sharpening (0.2)	0.9394	0.8462	0.8801	0.9969
JPEG (70) + Rotation (5°)	0.9964	0.8326	0.8677	0.9885
Scaling (2) + Sharpening (0.2)	0.9275	0.8306	0.9000	0.9959
Gaussian noise (0.02) + Median filter (3 × 3) + Sharpening (0.2)	0.8190	0.8320	0.8777	0.9925
Salt and Pepper noise (0.05) + Average filter (3 × 3) + Scaling (0.5)	0.9889	0.8306	0.8849	0.9936
Average filter (5 × 5) + Scaling (0.5) + JPEG (30)	0.9883	0.8402	0.8771	0.9980
Scaling (2) + Rotation (2°) + Gaussian filter (3 × 3)	0.9957	0.8339	0.9496	0.9960

As can be seen from Table 7, for JEPG compression, noise, filtering, sharpening, scaling, and most hybrid attacks, the NCC values of the proposed algorithm are higher than the other three algorithms, which means that the proposed algorithm is more robust than the other three algorithms. However, when the watermarked image is attacked by rotation attacks, the robustness of the proposed algorithm is slightly weaker than algorithm [18]. In general, SVD-based watermarking algorithms have stronger robustness than DCT-based watermarking algorithms, and the comparison results in Table 7 are no exception. When suffering noise attacks, sharpening attacks, and some hybrid attacks, the robustness of the proposed algorithm is much better than the other three algorithms.

4.4. Real-Time Analysis

To compare the real-time performance of the four algorithms, we take "Lena", "Baboon", "Airplane", "Sailboat", "House", and "Peppers" as objects to test the average execution time for watermark embedding and extraction. The test results are given in Table 8, including embedding time, extraction time, and total time. All values in Table 8 are obtained by averaging the execution time of six objects, and the execution time of each object is obtained by averaging 10 test results.

Table 8. Comparisons of the embedding time, extraction time, and total time among different algorithms (second).

Algorithms	Embedding Time	Extraction Time	Total Time
Patvardhan et al. [18]	1.4668	0.6477	2.1145
Su et al. [31]	5.0233	2.2093	7.2326
Jia et al. [34]	1.9622	0.4170	2.3792
Proposed	1.1560	0.3880	1.5440

From Table 8, we can see that the embedding time, extraction time, and total time of the proposed algorithm are all shorter than those of the other three algorithms. The execution time of reference [31] is the longest, and the execution time of reference [18] is similar to that of reference [34]. As for reference [31], its watermark preprocessing and embedding processes prolong the total execution time. In conclusion, the real-time performance of the proposed algorithm is better than the other three algorithms.

5. Conclusions

In this paper, a double color image watermarking algorithm is proposed. The main contributions are as follows. (1) The APDCBT transform is introduced into the color image watermarking algorithm. The better low-frequency energy accumulation characteristics of the APDCBT and the stability of SSVD are combined to ensure the robustness of the algorithm. (2) The Fibonacci transform with larger key space is used to improve the security of the algorithm. (3) The watermarks for embedding are color images, not gray or binary images. Additionally, the color QR code with error correction mechanism is introduced to be a watermark, which can further improve the robustness of the algorithm. The experimental results show that the algorithm can effectively resist common image processing attacks, such as JPEG compression, Gaussian noise, salt and pepper noise, average filter, median filter, Gaussian filter, sharpening, and scaling attacks, while its robustness to rotation attacks is relatively weak. In the future, we will consider how to improve the resistance of the algorithm to high-intensity rotation attacks.

Author Contributions: X.Y. and C.W. conceived the algorithm and designed the experiments; X.Y. performed the experiments; X.Y. and X.Z. analyzed the results; X.Y. drafted the manuscript; C.W. and X.Z. revised the manuscript. All authors read and approved the final manuscript.

Funding: This work was supported by the National Natural Science Foundation of China (No. 61702303), the Shandong Provincial Natural Science Foundation, China (No. ZR2017MF020), and the Science and Technology Development Plan Project of Weihai Municipality (No. 2018DXGJ07).

Conflicts of Interest: The authors declare no conflict of interest.

References

1. Zhang, X.Q.; Wang, X.S. Digital image encryption algorithm based on elliptic curve public cryptosystem. *IEEE Access* **2018**, *6*, 70025–70034. [CrossRef]
2. Denemark, T.; Boroumand, M.; Fridrich, J. Steganalysis features for content-adaptive JPEG steganography. *IEEE Trans. Inf. Forensics Secur.* **2016**, *11*, 1736–1746. [CrossRef]
3. Li, Y.N.; Wang, J.R. Robust content fingerprinting algorithm based on invariant and hierarchical generative model. *Digit. Signal Process.* **2019**, *85*, 41–53. [CrossRef]
4. Schyndel, R.G.V.; Tirkel, A.Z.; Osborne, C.F. A digital watermark. In Proceedings of the 1st IEEE International Conference on Image Processing, Austin, TX, USA, 13–16 November 1994.
5. Kamaruddin, N.S.; Kamsin, A.; Por, L.Y.; Rahman, H. A review of text watermarking: Theory, methods and applications. *IEEE Access* **2018**, *6*, 8011–8028. [CrossRef]
6. Wang, C.Y.; Zhang, Y.P.; Zhou, X. Robust image watermarking algorithm based on ASIFT against geometric attacks. *Appl. Sci.* **2018**, *8*, 410. [CrossRef]

7. Abdulfetah, A.A.; Sun, X.M.; Yang, H.F.; Mohammad, N. Robust adaptive image watermarking using visual models in DWT and DCT domain. *Inf. Technol. J.* **2010**, *9*, 460–466. [CrossRef]
8. Gupta, R.; Mishra, A.; Jain, S. A semi-blind HVS based image watermarking scheme using elliptic curve cryptography. *Multimed. Tools Appl.* **2018**, *77*, 19235–19260. [CrossRef]
9. Erkucuk, S.; Krishnan, S.; Zeytinoglu, M. A robust audio watermark representation based on linear chirps. *IEEE Trans. Multimed.* **2006**, *8*, 925–936. [CrossRef]
10. Mohammed, A.A.; Ali, N.A. Robust video watermarking scheme using high efficiency video coding attack. *Multimed. Tools Appl.* **2018**, *77*, 2791–2806. [CrossRef]
11. Sathya, S.P.A.; Ramakrishnan, S. Fibonacci based key frame selection and scrambling for video watermarking in DWT–SVD domain. *Wireless Pers. Commun.* **2018**, *102*, 2011–2031. [CrossRef]
12. Himeur, Y.; Boukabou, A. A robust and secure key-frames based video watermarking system using chaotic encryption. *Multimedia Tools Appl.* **2018**, *77*, 8603–8627. [CrossRef]
13. Chou, C.H.; Wu, T.L. Embedding color watermarks in color images. *EURASIP J. Appl. Signal Process.* **2003**, *1*, 32–40. [CrossRef]
14. Su, Q.T.; Yuan, Z.H.; Liu, D.C. An approximate Schur decomposition-based spatial domain color image watermarking method. *IEEE Access* **2019**, *7*, 4358–4370. [CrossRef]
15. Su, Q.T.; Liu, D.C.; Yuan, Z.H.; Wang, G.; Zhang, X.F.; Chen, B.J.; Yao, T. New rapid and robust color image watermarking technique in spatial domain. *IEEE Access* **2019**, *7*, 30398–30409. [CrossRef]
16. Barni, M.; Bartolini, F.; Piva, A. Multichannel watermarking of color image. *IEEE Trans. Circuits Syst. Video Technol.* **2002**, *12*, 142–156. [CrossRef]
17. Liu, K.C. Wavelet-based watermarking for color images through visual masking. *AEU - Int. J. Electron. Commun.* **2010**, *64*, 112–124. [CrossRef]
18. Patvardhan, C.; Kumar, P.; Lakshmi, C.V. Effective color image watermarking scheme using YCbCr color space and QR code. *Multimed. Tools Appl.* **2018**, *77*, 12655–12677. [CrossRef]
19. Roy, S.; Pal, A.K. A blind DCT based color watermarking algorithm for embedding multiple watermarks. *Aeu-Int. J. Electron. Commun.* **2017**, *72*, 149–161. [CrossRef]
20. Niu, P.P.; Wang, X.Y.; Liu, Y.N.; Yang, H.Y. A robust color image watermarking using local invariant significant bitplane histogram. *Multimed. Tools Appl.* **2017**, *76*, 3403–3433.
21. Liu, X.Y.; Wang, Y.F.; Du, J.Y.; Liao, S.H.; Lou, J.T.; Zou, B.J. Robust hybrid image watermarking scheme based on KAZE features and IWT-SVD. *Multimed. Tools Appl.* **2019**, *78*, 6355–6384. [CrossRef]
22. Chang, T.J.; Pan, I.H.; Huang, P.S.; Hu, C.H. A robust DCT-2DLDA watermark for color images. *Multimed. Tools Appl.* **2019**, *78*, 9169–9191. [CrossRef]
23. Lakrissi, Y.; Saaidi, A.; Essahlaoui, A. Novel dynamic color image watermarking based on DWT-SVD and the human visual system. *Multimed. Tools Appl.* **2018**, *77*, 13531–13555. [CrossRef]
24. Li, J.Z.; Yu, C.Y.; Gupta, B.B.; Ren, X.C. Color image watermarking scheme based on quaternion Hadamard transform and Schur decomposition. *Multimed. Tools Appl.* **2018**, *77*, 4545–4561. [CrossRef]
25. Rosales-Roldan, L.; Chao, J.H.; Nakano-Miyatake, M.; Perez-Meana, H. Color image ownership protection based on spectral domain watermarking using QR codes and QIM. *Multimed. Tools Appl.* **2018**, *77*, 16031–16052. [CrossRef]
26. Roy, S.; Pal, A.K. A hybrid domain color image watermarking based on DWT-SVD. *Iran. J. Sci. Technol.-Trans. Electr. Eng.* **2019**, *43*, 201–217. [CrossRef]
27. Vaidya, P.S.; Mouli, C.P.V.S.S.R. A robust semi-blind watermarking for color images based on multiple decompositions. *Multimed. Tools Appl.* **2017**, *76*, 25623–25656.
28. Mohammad, M.; Gholamhossein, E. A new DCT-based robust image watermarking method using teaching-learning-based optimization. *J. Inf. Secur. Appl.* **2019**, *47*, 28–38.
29. Laur, L.; Rasti, P.; Agoyi, M.; Anbarjafari, G. A robust color image watermarking scheme using entropy and QR decomposition. *Radio Eng.* **2015**, *24*, 1025–1032. [CrossRef]
30. Cedillo-Hernandez, M.; Cedillo-Hernandez, A.; Garcia-Ugalde, F.J.; Nakano-Miyatake, M.; Perez-Meana, H. Copyright protection of color imaging using robust-encoded watermarking. *Radio Eng.* **2015**, *24*, 240–251. [CrossRef]
31. Su, Q.T.; Wang, G.; Jia, S.L.; Zhang, X.F.; Liu, Q.M.; Liu, X.X. Embedding color image watermark in color image based on two-level DCT. *Signal Image Video Process.* **2015**, *9*, 991–1007. [CrossRef]

32. Su, Q.T.; Chen, B.J. An improved color image watermarking scheme based on Schur decomposition. *Multimed. Tools Appl.* **2017**, *76*, 24221–24249. [CrossRef]
33. Pandey, M.K.; Parmar, G.; Gupta, R.; Sikander, A. Non-blind Arnold scrambled hybrid image watermarking in YCbCr color space. *Microsyst. Technol.* **2019**, *25*, 3071–3081. [CrossRef]
34. Jia, S.L.; Zhou, Q.P.; Zou, H. A novel color image watermarking scheme based on DWT and QR decomposition. *J. Appl. Sci. Eng.* **2017**, *20*, 193–200.
35. Hou, Z.X.; Wang, C.Y.; Yang, A.P. All phase biorthogonal transform and its application in JPEG-like image compression. *Signal Process. Image Commun.* **2009**, *24*, 791–802. [CrossRef]
36. Zhou, X.; Zhang, H.; Wang, C.Y. A robust image watermarking technique based on DWT, APDCBT, and SVD. *Symmetry* **2018**, *10*, 77. [CrossRef]
37. Zhang, Z.; Wang, C.Y.; Zhou, X. Image watermarking scheme based on DWT-DCT and SSVD. *Int. J. Secur. Its Appl.* **2016**, *10*, 191–205.
38. Cai, Y.J.; Niu, Y.G.; Su, Q.T. Blind watermarking algorithm for color images based on DWT-SVD and Fibonacci transformation. *Appl. Res. Comput.* **2012**, *29*, 3025–3028.
39. University of Granada, Computer Vision Group: CVG-UGR Image Database. Available online: http://decsai.ugr.es/cvg/dbimagenes/c512.php (accessed on 12 July 2019).
40. Scholar Homepage of Chengyou Wang. Available online: http://www.scholat.com/wangchengyou (accessed on 15 July 2019).
41. Ladislav, P.; Jan, K.; Ondrej, Z.; Ondrej, K.; Libor, B.; Martin, S.; Tomas, K. Study of advanced compression tools for stereoscopic video by objective metrics. In Proceedings of the 26th International Conference Radioelektronika, Kosice, Slovakia, 19–20 April 2016.

Article

Exploration with Multiple Random ε-Buffers in Off-Policy Deep Reinforcement Learning

Chayoung Kim [1] and JiSu Park [2,*]

[1] Division of General Studies, Kyonggi University, Suwon, Gyeonggi-do 16227, Korea; kimcha0@kgu.ac.kr
[2] Convergence Institute, Dongguk University, Seoul 04620, Korea
* Correspondence: bluejisu@dgu.edu; Tel.: +82-2-2290-1491

Received: 26 September 2019; Accepted: 30 October 2019; Published: 1 November 2019

Abstract: In terms of deep reinforcement learning (RL), exploration is highly significant in achieving better generalization. In benchmark studies, ε-greedy random actions have been used to encourage exploration and prevent over-fitting, thereby improving generalization. Deep RL with random ε-greedy policies, such as deep Q-networks (DQNs), can demonstrate efficient exploration behavior. A random ε-greedy policy exploits additional replay buffers in an environment of sparse and binary rewards, such as in the real-time online detection of network securities by verifying whether the network is "normal or anomalous." Prior studies have illustrated that a prioritized replay memory attributed to a complex temporal difference error provides superior theoretical results. However, another implementation illustrated that in certain environments, the prioritized replay memory is not superior to the randomly-selected buffers of random ε-greedy policy. Moreover, a key challenge of hindsight experience replay inspires our objective by using additional buffers corresponding to each different goal. Therefore, we attempt to exploit multiple random ε-greedy buffers to enhance explorations for a more near-perfect generalization with one original goal in off-policy RL. We demonstrate the benefit of off-policy learning from our method through an experimental comparison of DQN and a deep deterministic policy gradient in terms of discrete action, as well as continuous control for complete symmetric environments.

Keywords: deep Q-network (DQN); reinforcement learning (RL); explorations; deep deterministic policy gradient (DDPG); random ε-greedy buffers

---

1. Introduction

Deep reinforcement learning (RL) [1] has been applied in challenging domains, such as games and robotics. A combination of RL and nonlinear function approximates, such as deep neural networks, helps in automating decision making and control issues. However, it can also be challenging in terms of stability and convergence [2] in real-time online learning, such as in the detection of network security through the verification of whether the network is "normal or anomalous" [3–7]. Moreover, its widespread adaptation in the real world, such as with robotic arms, is difficult because of sample complexity and heavy dependence on certain hyper-parameters, such as exploration constants [3–7]. A significant challenge in robotics is the engineering of a carefully shaped reward function. However, the necessity of reward function engineering restricts the general applicability of RL in the real world. Recent remarkable research on hindsight experience replay (HER) [8] provides additional replay buffers with the original goal and a subset of other goals without any domain knowledge. HER improves the sample-efficient setting and learns the possible multi-goals, even in an environment of sparse rewards. Off-policy algorithms in RL [1], such as deep Q-networks (DQNs) [9,10], reuse the past experience with random ε-greedy actions in replay buffers for efficient exploration behaviors. Andrychowicz et al. [8] found that a DQN combined with HER resolves some tasks more easily than that without

HER. An experience comprises a tuple, such as state, action, reward, or new state, and each agent of DQN can exploit a batch of experiences to update the Q-value function via temporal differencing (TD) learning [9,10]. The experience replay (ER) buffer discards old memories at each step by sampling the buffer randomly and updating the DQN agent. It aids in breaking temporal relations and increasing data usage [11,12].

The present study is inspired by the quantification of generalization in RL [11], which investigates the impact of injecting stochastic methods on generalization by random ε-greedy actions overriding the agent's preferred actions. A random ε-greedy action has been used to encourage exploration and prevent excessive over-fitting [11]. There has been research on the exploration diversities of stability and convergence. A prioritized replay memory [13] has been used in a DQN for replaying important transitions more frequently and learning more efficiently. HER [8] makes possible sample-efficient learning with additional replay buffers. A parameter space in noise [14] in a neural network leads to a more abundant set of exploratory behaviors through noise injection in the action space in deep RL. However, some empirical implementations show that in certain environments, adding parameter space noise [14] may not improve ε-greedy random actions by OpenReivew.net [15]. A prioritized replay memory attributed to a complex TD error might provide better theoretical results. The TD errors might create a balance issue between variance and bias, because TD attributes can be computed at every n-time step [11,12]. Moreover, some studies have only partially succeeded in combining drop-out, L2 regularization, and ε-greedy random actions. Thus, training with only ε-greedy random actions can vastly improve generalization and exploration [11,12].

This paper proposes multiple ε-greedy experience buffers in off-policy deep RL [1] for enhancing explorations for a more near-perfect generalization for the original goal. We consider strengthening the advantages of exploring model-free, deep RL, and demonstrate that the off-policy method benefits through the experimental comparison of DQN [9,10] and DDPG [16], which is actually based on on-policy, and represents a trade-off between policy optimization [2,16] and Q-learning [9,10]. The proposed deep RL model is compatible with the discrete action as well as continuous control, symmetrically. Therefore, we can expect better prediction accuracy in real-time online learning, such as detecting network intrusions by verifying whether the network is "normal or anomalous" [3–7]. The concept of our proposed multiple ε-buffers is extremely simple. After experiencing some episodes, we store them in the replay buffers R_1 and R_2. Note that we can replay one trajectory with only one goal, assuming that we exploit an off-policy RL like DQN [9,10] and DDPG [16]. When the procedure passes through multiple ε-buffers, the trade-off between policy optimization and Q-learning can be solid and strong due to the deep neural network in DQN [9,10] or DDPG [16].

2. Background

RL comprises an agent with an environment that is described by a set of states, S, a set of actions, A, a distribution of initial states, $p(s_0)$, a reward function, r: $S \times A \rightarrow R$, transition probabilities, $p(s_{t+1}|s_t,a_t)$, and a discount factor, $\gamma \in [0,1]$ [1]. Deterministic policy maps from states to actions are given as π: $S \rightarrow A$. At each time step, t, the agent selects an action attributed to the current state: $a_t = \pi(s_t)$, and takes the reward, $r_t = r(s_t,a_t)$ [1]. The new state of the environment is sampled from the distribution $p(|s_t,a_t)$. A discounted cumulative reward is a return, $R_t = \sum T\gamma^{t'-t}r_{t'}$, where T is the time step at which the agent's simulation terminates [1]. The goal here is to maximize its expected return $E_{s0}[R_0|s_0]$. The action value function, named Q-function, is defined as $Q^{\pi}(s_t,a_t) = E[R_t|s_t,a_t]$. An optimal policy is denoted by the following equation, called the Bellman equation, $Q^*(s,a) = E_{s'\sim p(|s,a)}\,[r + \gamma \max_{a'} Q^*(s',a')|s,a]$, (1), this equation converges at the Q-function as an optimal action value [1]. The Bellman equation is to be used as a function approximation for estimating the Q-function, $Q(s,a;\theta) \approx Q^*(s,a)$, as an action value [1]. This can be a linear-function approximation, but it is sometimes even a nonlinear-function approximation, such as with a deep neural network [9,10]. Therefore, we attempt to exploit a deep neural network as a function approximation. We refer to a deep neural network approximation with weights θ as the Q-function [9,10]. A Q-function network can be trained by minimizing a sequence of loss functions, $L_i(\theta_i)$, which changes at i per iteration [1,9,10]. $L_i(\theta_i) = E_{s,a\sim p(|s,a)}[(y_i - Q(s,a;\theta_i))^2]$, (2), where $y_i = E_{s'\sim p(|s,a)}$ $[r + \gamma max_{a'}Q^*(s',a';\theta_{i-1})|s,a]$ is the target for i per iteration. The loss function can be differentiated with regard to the weights, depending on the gradient [1,9,10]. $\nabla\theta_i L_i(\theta_i) = E_{s,a\sim p(|s,a);s'\sim\varepsilon}[r + \gamma max_{a'}Q(s',a';\theta_{i-1}) - Q(s,a;\theta_i))\nabla\theta_i Q(s,a;\theta_i)]$, (3), [1,9,10]. Optimizing the loss function by using the stochastic gradient descent is computationally intricate. Q-learning as a function approximation becomes possible [17,18], and is model-free, which indicates the RL model and off-policy. This implies that it learns about greedy strategy $a = max_a Q(s,a;\theta)$ while obeying an exploration selected by ε-greedy with a probability of 1 − ε and a random action with a probability of ε for exploration and exploitation [17,18]. DQN is a well-known, model-free RL algorithm attributed to discrete action spaces. In DQN [9,10], we construct a deep neural network, Q, which approximates Q* and is greedy-defined as $\pi_Q(s) = argmax_{a\in A}Q(s,a)$ [9,10]. It is a ε-greedy policy with probability ε and takes the action $\pi_Q(s)$ with probability 1 − ε. Each episode uses the ε-greedy policy following Q as an approximation of the deep neural network. The tuples (s_t, a_t, r_t, s_{t+1}) are stored in the replay buffer, and each new episode is configured to neural network training [9,10]. The deep neural network is trained using the gradient descent of random episodes on loss L, encouraging Q to follow the Bellman equation [9,10]. The tuples are sampled from the replay buffer of random episodes. The target network y_t is computed by a separate neural network that changes more slowly than the main deep neural network in order to optimize a stable process. The weights of the target network are set to the current weights of the main deep neural network. The DQN algorithm [9,10] is presented in Algorithm 1.

Algorithm 1 Deep Q-learning with Experience Replay

Initialize replay memory D to capacity N
Initialize action-value function Q with random weights
for episode = 1, *M* do
  Initialize sequence $s_1 = \{x_1\}$ and preprocessed sequenced $\phi_1 = \phi(s_1)$
  for t = 1, *T* do
    With probability ε, select a random action α_t
    Otherwise select $\alpha_t = max_\alpha Q^*(\phi(s_t),\alpha;\theta)$
    Execute action α_t in emulator and observe reward r_t and image x_{t+1}
    Set $s_{t+1} = s_t, \alpha_t, x_{t+1}$ and preprocess $\phi_{t+1} = \phi(s_{t+1})$
    Store transition $(\phi_t, \alpha_t, r_t, \phi_{t+1})$ in D
    Sample ***random mini-batch of transitions*** $(\phi_j, \alpha_j, r_j, \phi_{j+1})$ *from* D
    Set $y_j = r_j$ for terminal ϕ_{j+1}
    Or $y_j = r_j + \gamma max_{\alpha'}Q^*(\phi_{j+1},\alpha';\theta)$ for non-terminal ϕ_{j+1}
    Perform a gradient descent step on $(y_j - Q(\phi_j, \alpha_j|\theta))^2$
  end for
end for

A deep deterministic policy gradient (DDPG) [16] is a well-known, model-free RL algorithm attributed to continuous action spaces [2,16]. In a DDPG, we construct two neural networks: a target policy as an actor, π: S → A, and an action-value-function approximation as a critic, Q: S × A → R. The critic approximates Q^π, which is the action-value function of the actor [16]. Each episode uses a noisy policy of the target, $\pi_b(s) = \pi(s) + N(0,1)$ [16]. The critic is trained in the same way as the Q-function in a DQN [9,10]. The target network yt is computed using the output of the actor, i.e., $y_t = r_t + \gamma Q(s_{t+1}, \pi(s_{t+1}))$, and the actor's deep neural network is trained using the gradient descent of random episodes on the loss $L_a = -E_s Q(s,\pi(s))$ [2,16] and sampled from the replay buffer of random episodes. The gradient of L_a is computed by both the critic and actor through back-propagation [2,16]. The algorithm of a DDPG [16] is shown in Algorithm 2.

Algorithm 2 Deep Deterministic Policy Gradient

Randomly initialize critic $Q(s, a|\theta^Q)$ and actor $\mu(s|\theta^\mu)$ with weights θ^Q and θ^μ
Initialize target Q′ and μ′ with weights $\theta^{Q'} \leftarrow \theta^Q$, $\theta^{\mu'} \leftarrow \theta^\mu$
Initialize replay memory R
for episode = 1, *M* do
    Initialize a random process N for action exploration
    Receive initial observation state s_1
    for t = 1, *T* do
        Select action $\alpha_t = \mu(s_t|\theta^\mu) + \mathcal{N}_t$ according to the current policy and exploration noise
        Execute action α_t and observe reward r_t and new state s_{t+1}
        Store transition $(s_t, \alpha_t, r_t, s_{t+1})$ in R
        Sample ***random mini-batch of transitions*** $(s_j, \alpha_j, r_j, s_{j+1})$ from R
        Set $y_j = r_j + \gamma Q'(s_{j+1}, \mu'(s_{j+1}|\theta^{\mu'})|\theta^{Q'})$
        Update critic by minimizing the loss: $L = \frac{1}{n}\Sigma_j(y_j - Q(s_j, \alpha_j|\theta^Q))^2$
        Update the actor policy using the sampled policy gradient:
            $\nabla_{\theta\mu} I \approx \frac{1}{n}\Sigma_j \nabla_\alpha Q(s,\alpha|\theta^Q)|s = s_j, \alpha = \alpha\ \mu(s_j)\nabla_{\theta\mu}\mu(s|\theta^\mu)|s_j$
        Update the target:
            $\theta^{Q'} \leftarrow \tau\theta^Q + (1-\tau)\ \theta^{Q'}$
            $\theta^{\mu'} \leftarrow \tau\theta^\mu + (1-\tau)\ \theta^{\mu'}$
    end for
end for

3. Multiple Random ε-Buffers

3.1. Proposed Off-Policy Algorithm

We propose multiple random ε-greedy buffers in a DQN [9,10] for discrete action spaces and DDPG [2,16] for continuous action spaces. The idea of our proposed multiple ε-buffers is to store experienced episodes in the replay buffers R_1 and R_2. For better exploration, the agent evaluates a target policy $\pi(a|s)$ to compute $V\pi$ (s) or $Q\pi(s,a)$ while following a behavior policy $\mu(a|s)$, where $\{s_1, a_1, r_2, \ldots s_T\} \sim \mu$. The agent learns about multiple policies while following one policy. This means that it can learn about the optimal policy and follow the exploratory policy. The target policy π is greedy with respect to Q(s,a), such as $\pi(s_{t+1}) = \text{argmax}_{a'} Q(s_{t+1},a')$, while the behavior policy μ is ε-greedy with respect to Q(s,a), such as $\mu(s_{t+1}) = R_{t+1} + \max_{a'} \gamma Q(s_{t+1},a')$. When an agent follows a greedy policy, convergence changes rapidly. However, because of a lack of exploration, the agent reaches local optimization. Many other challenges address this global–local dilemma [2,9,10,16]. A remarkable recent work, HER [8], provides additional replay buffers with the original goal and a subset of other goals without any complicated reward function. The off-policy algorithms in RL, such as DQN [9,10] or DDPG [2,16], reuse the past experience with random ε-greedy actions in replay buffers for better exploration. The ER buffers discard old memories at each step by sampling the buffer randomly to update the DQN agent. Therefore, ER helps to break temporal relations and increase data usage [11,12].

Our study is inspired by quantifying generalization in RL [11], which provides a random ε-greedy action that has been used to encourage exploration and prevent excessive over-fitting. A prioritized replay memory [13] has been exploited in a DQN to replay important transitions. A parameter space in the noise [14] of a neural network leads to a more abundant set of exploratory behaviors. These recent studies have shown how HER [8] can improve random ε-greedy actions. Therefore, we consider enhancing multiple random ε-greedy experience buffers in deep RL for more near-perfect explorations for the original goal. We demonstrate the off-policy learning benefit from our method through an experimental comparison of DQN [9,10] and DDPG [2,16] in high-dimensional discrete action environments as well as continuous control tasks. Our idea is that after experiencing some episodes, we store them in the replay buffers R_1 and R_2. We can replay one trajectory with only one goal, assuming that we exploit an off-policy RL like DQN [9,10] and DDPG [2,16]. While the procedure passes through multiple random ε-buffers, the balance between variance and bias of the policy can be solid and strong due to the training of the deep neural network.

Without using a complicated shaping reward and knowledge environment, we attempt to achieve a near-perfect global optimal goal. This idea is also used in the HER [8] and ODE model [12]. We attempt to achieve better exploration, which can balance computational speed and global optimization. If we use both a target policy and an alternative policy, we can attain the previously achieved global optimization. Subsequently, we consider the computational speed while maintaining generalization. HER [8] stores each transition in the replay buffer not only with the original goal, but also with a subset of other goals. However, we increase the number of replay buffers, which are a minimum of two. Therefore, we can replay each trajectory with two or more buffers, assuming that we use an off-policy RL algorithm, such as DQN [9,10] or DDPG [2,16], for a complete symmetric environment. Similar to HER [8], our proposed model can learn with extremely sparse rewards, and performs better in our experiments. These results indicate the practical challenges with reward shaping, such as HER [8]. Moreover, we can acquire the results of the effects of memory size from ODEs [12]. The selection of memory size does not have any effect on learning behavior; there is a trade-off between an over-fitting and the weight update [12]. If the memory size is smaller, the process of learning is probably over-fitted. Similarly, if the memory is bigger, the over-fitting effect is alleviated. However, it is not easy to deal with the balance between over-fitting and weight update [11,12]. With prioritized replay [13], if the memory size is small, especially if the mini-batch is small, then the trade-off between the over-fitting and weight update is more consequential. Therefore, we consider enhancing the characteristic feature of model-free RL, i.e., increasing the number of random ε-buffers. In terms of policy optimization, the policy $\pi\theta(s,a)$ is explicit, and for performance optimization, the agent follows the policy $\pi\theta(s,a)$ directly. Therefore, we call it on-policy optimization. However, Q-learning is based on off-policy, because it can learn about an optimal policy with respect to Q(s,a), such as $\pi(s_{t+1}) = \text{argmax}_{a'}Q(s_{t+1},a')$, and follow an exploratory policy [1]. Policy optimization [2,16] is more stable than Q-learning [9,10]; however, Q-learning [9,10] is more sample-efficient than policy optimization [2,16]. Therefore, we follow the lead to strengthen the advances in model-free RL. DDPG [16] is actually based on on-policy [1]. However, it can create a good trade-off between policy optimization [2,16] and Q-learning [9,10]. Algorithms 3 and 4 display the DQN and DDPG algorithm, respectively, with multiple random ε-buffers.

Algorithm 3 Deep Q-learning with multiple random ε-buffers

Initialize replay memory R_1 and R_2
Initialize Q-function with random weights
for episode = 1, M do
    Initialize sequence s_1 and $\phi_1 = \phi(s_1)$
    for t = 1, T do
        With probability ε, select a random action α_t
            Or select $\alpha_t = \max_\alpha Q^*((s_t), \alpha;\theta)$
        Execute action α_t in emulator and observe reward r_t
        Set $s_{t+1} = s_t, \alpha_t$ and $\phi_{t+1} = \phi(s_{t+1})$
          Store transition $(\phi_t, \alpha_t, r_t, \phi_{t+1})$ ***in* $\mathbf{R_1}$**
                //standard experience buffer
          Store transition $(\phi_t, \alpha_t, r_t, \phi_{t+1})$ ***in* $\mathbf{R_2}$**
                //second experience buffer
        Sample ***random mini-batch of transitions*** $(\phi_j, \alpha_j, r_j, \phi_{j+1})$ ***from either* $\mathbf{R_1}$ *or* $\mathbf{R_2}$**
          Set $y_j = r_j$
            Or $y_j = r_j + \gamma\max_{\alpha'}Q^*(\phi_{j+1}, \alpha'|\theta)$
        Perform a gradient descent step on $(y_j - Q(\phi_j, \alpha_j|\theta))^2$
    end for
        Sample ***random mini-batch of transitions from either* $\mathbf{R_1}$ *or* $\mathbf{R_2}$**
end for

Algorithm 4 Deep Deterministic Policy Gradient with multiple random ε-buffers

Initialize critic $Q(s, a|\theta^Q)$ and actor $\mu(s|\theta^\mu)$ with weights θ^Q and θ^μ
Initialize target Q' and μ' with weights $\theta^{Q'} \leftarrow \theta^Q, \theta^{\mu'} \leftarrow \theta^\mu$
Initialize replay memory R_1 and R_2
for episode = 1, *M* do
  Initialize observation s_1 and a random process *N* for action exploration
  for t = 1, *T* do
    Select action $\alpha_t = \mu(s_t|\theta^\mu) + \mathcal{N}_t$ by the current policy and exploration noise
    Execute action α_t and observe reward r_t and new state s_{t+1}
      Store transition $(s_t, \alpha_t, r_t, s_{t+1})$ ***in* $\mathbf{R_1}$**
        //standard experience buffer
      Store transition $(s_t, \alpha_t, r_t, s_{t+1})$ ***in* $\mathbf{R_2}$**
        //second experience buffer
    Sample ***a random mini-batch of transitions*** $(s_j, \alpha_j, r_j, s_{j+1})$ ***from either* $\mathbf{R_1}$ *or* $\mathbf{R_2}$**
      Set $y_j = r_j + \gamma Q'(s_{j+1}, \mu'(s_{j+1}|\theta^{\mu'})|\theta^{Q'})$
      Update critic by minimizing the loss: $L = \frac{1}{n}\Sigma_j(y_j - Q(s_j, \alpha_j|\theta^Q))^2$
    Update the actor by the sampled policy gradient:
      $\nabla_{\theta^\mu} I \approx \frac{1}{n}\Sigma_j \nabla_\alpha Q(s,\alpha|\theta^Q)|s = s_j, \alpha = \alpha\ \mu(s_j)\nabla_{\theta^\mu}\mu(s|\theta^\mu)|s_j$
    Update the target:
      $\theta^{Q'} \leftarrow \tau\theta^Q + (1-\tau)\,\theta^{Q'}$
      $\theta^{\mu'} \leftarrow \tau\theta^\mu + (1-\tau)\,\theta^{\mu'}$
  end for
    Sample ***a random mini-batch of transitions from either* $\mathbf{R_1}$ *or* $\mathbf{R_2}$**
end for

3.2. *Algorithm Description*

3.2.1. DQN with Multiple Random ε-Buffers

1. Initialize replay memories of both R_1 and R_2 and Q(s,a), and initiate the process from a random state, s.
2. Initialize the sequence with the start state, s.
3. The agent learns the policy $\max_\alpha Q^*((s_t, a);\theta)$ with greed and follows another policy with probability ε.
4. For exploration, an experience composed of a tuple, such as (state s, action a, reward r, new state s'), is in both R_1 and R_2.
5. Sample a random mini-batch of transitions from either R_1 or R_2.
6. The weights for performing the gradient descent $(r_j + \gamma \max_{\alpha'} Q^*(\phi_{j+1}, \alpha'|) - Q(\phi_j, \alpha_j|\theta))$ for a target DQN with multiple random ε-buffers.
7. Steps 3–6 are repeated for training.
8. For the next episode, sample a random mini-batch of transitions from either R_1 or R_2.
9. Steps 2–8 are repeated for training.

3.2.2. DDPG with Multiple Random ε-Buffers

1. Initialize critic deep network $Q(s, a|\theta^Q)$ and actor deep network $\mu(s|\theta^\mu)$.
2. Initialize replay memories in both R_1 and R_2.
3. The agent selects the action according to the current policy, $\alpha_t = \mu(s_t|\theta^\mu) + \mathcal{N}_t$.
4. For exploration, an experience composed of a tuple, such as (state s, action a, reward r, new state s'), is in both R_1 and R_2.
5. A mini-batch of transitions is randomly sampled from either R_1 or R_2.
6. The critic network is updated by the loss function $L = 1/n\sum_j(r_j + \gamma Q'(s_{j+1}, \mu'(s_{j+1}|\theta^{\mu'})|\theta^{Q'}) - Q(s_j, \alpha_j|\theta^Q))^2$.

7. The actor network is updated by the policy gradient descent $\nabla\theta\mu I \approx 1/n\sum_j \nabla_\alpha Q(s|\theta^Q)|s = s_j$, $\alpha = \alpha$ $\mu(s_j)$ $\nabla_{\theta\mu}\mu(s|\theta^\mu)|s_j$ for a target DDPG with multiple random ε-buffers.
8. Update the target network $\theta^{Q'} \leftarrow \tau\theta^Q + (1 - \tau)\,\theta^{Q'}$ and $\theta^{\mu'} \leftarrow \tau\theta^\mu + (1 - \tau)\,\theta^{\mu'}$.
9. Steps 3–8 are repeated for training.
10. For the next episode, a mini-batch of transitions is randomly sampled from either R_1 or R_2.
11. Steps 2–10 are repeated for training.

4. Evaluation and Results

We consider OpenAI Gym [19] for our proposed algorithms, i.e., DQN and DDPG with multiple random ε-buffers. To enhance multiple random ε-greedy experience buffers, we exploit classic control environments in OpenAI Gym, such as CartPole-v0 [20,21], MountainCar-v0 [22,23], and Pendulum-v0 [24,25]. With similar approaches [9,10], we exploit a method, known as experience replay. Moreover, in our proposed method, Q-learning stores past experiences at each time step, $e_t = (s_t, a_t, r_t, s_{t+1})$, in multiple random ε-buffers, R_1 and R_2. Our emulators from OpenAI Gym [19] can apply mini-batch updates into R_1 and R_2. After the multiple random experience replay memories, the agent's actions of the emulator follow ε-greedy policy. In terms of DQN with multiple random ε-buffers, for enhanced exploration, we also follow the theory of the target Q-network of breakthrough research in [9,10]. The Q-learning agent calculates the TD error with the current estimated Q-value [9,10]. The optimized action-value function follows a significant identity, known as the Bellman equation [9,10]. In accordance with this equation, the TD target is the reward based on an action in the state plus the discounted highest Q-value for the next state [9,10]. In terms of DDPG with multiple random ε-buffers, for enhanced exploration, we also follow the theory of the critic and actor networks in [2,16]. Because the Q-learning of DDPG can exploit the deterministic policy, $\text{argmax}_a Q$, our proposed method can follow the off-policy [1].

For DQN with multiple random ε-buffers, we exploit CartPole-V0 [20,21] and MountainCar-V0 [22,23]. In CartPole-V0 [20,21], a pole is attached with an unactuated joint to a cart moving along a frictionless track. It is controlled by forcing +1 or −1 to the cart. The pole starts upright, and the goal is to prevent it from falling over. A + 1 reward is given to every time step in which the pole remains upright. The episode stops when the pole is more than 15° from the vertical direction or the cart moves more than 2.4 units from the center. In our simulation, CartPole-V0 defines "solving" as if the average reward is more than 490 or equal to 500 over 10 consecutive runs. The agent of CartPole-V0 receives −100 reward if it falls over prior to the max-length of the episode [21]. In MountainCar-V0, a car is on a one-dimensional track, positioned between two mountains, with the goal of driving up the mountain on the right. The engine of the car is not strong enough to scale the mountain in a single pass. Therefore, the only way to succeed is to drive back-and-forth to reach the other side of the mountain. MountainCar-V0 defines "solving" as obtaining an average reward of −110 over 100 consecutive trails [23]. In Pendulum-V0, the inverted pendulum starts from a random position, and the goal is to swing it upwards so that it stays upright. Pendulum-V0 is an unsolved environment, which means that it does not have a special reward threshold at which it is considered solved [24,25].

Our proposed DQN and DDPG with multiple random ε-buffers were implemented using TensorFlow [26] and Keras [27] in OpenAI Gym [19].

4.1. CartPole-V0

For DQN with multiple random ε-buffers, we follow similar previous studies [21]. In Figure 1a, there is a CartPole-V0 [20] by OpenAI Gym [19]. Thus, GAMMA = 0.95, LEARNING − RATE = 0.001, MEMORY − SIZE = 10,000, BATCH − SIZE = 64, EXPLORATION − MAXIMUM = 1.0, EXPLORATION − MININUM = 0.01, and EXPLORATION − DECAY = 0.995 are the same as with the previous DQN studies [21] because of a fairness comparison.

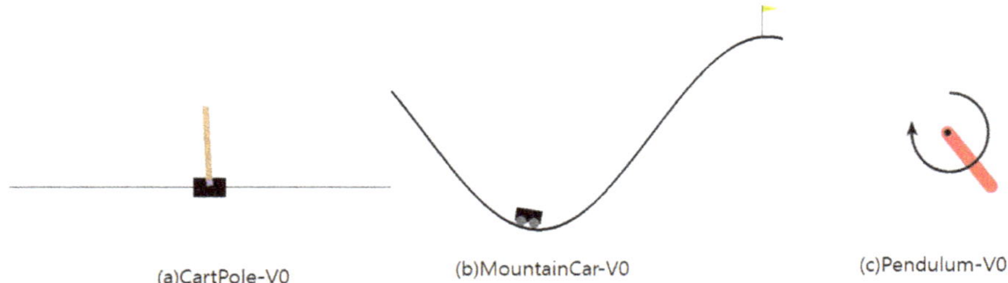

Figure 1. (**a**) CartPole-V0 [20], (**b**) MountainCar-V0 [22], and (**c**) Pendulum-V0 [24].

In the Cart-Pole-V0 environment, there are four observations and two discrete actions; see Figure 2. Furthermore, there are three episode terminations: the pole angle is more than −12° or + 12°, the cart position is more than −2.4 or + 2.4, and the episode length is greater than 200 [20].

(a) Environment Observation

Number	Observation	Minimum	Maximum
0	Cart Position	-2.4	+2.4
1	Cart Velocity	-Inf	+Inf
2	Pole Angle	~-41.8°	~+41.8°
3	Pole Velocity At Tip	-Inf	+Inf

(b) Actions

Number	Action
0	Push Cart to the Left
1	Push Cart to the Right

Figure 2. (**a**) The environment of CartPole-V0 and (**b**) the actions taken by the agent of CartPole-V0 [20].

However, the requirements from the implementation research can be considered in order to obtain better solutions [21]. Q-learning receives −100 as the reward when it falls before the maximum length of the episode is reached. Moreover, if the average reward is more than 490 or equal to 500 over 10 consecutive runs, Q-learning is terminated, even before it attains the maximum length of the episode [21].

Figures 3–5 display most of the results for the average, worst, and best cases, respectively. A DQN with multiple random ε-buffers can yield better results than one random buffer. In each of the average, worst, and best cases, our proposed method reaches the maximum reward earlier than DQN with one random buffer. Therefore, we can simulate identical results in real-time online for information security in Big Data. If our proposed method is used in training in real-time online in deep RL, as in a recent study [3–7], we acquire better results. Moreover, we can attempt a different deep neural network and hyper-parameter settings to demonstrate that our proposed model can enhance the exploration without specific domain information. We are convinced that we can enhance the behavior policy. Therefore, our proposed method can advance the exploration performance of ε-greedy, because multiple mini-batches do not follow a local minimum [9,10].

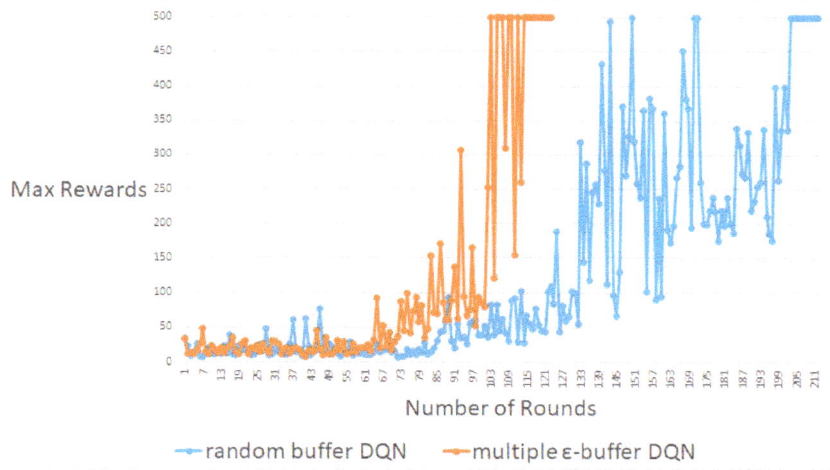

Figure 3. An average result between DQN with random buffers and DQN with multiple random ε-buffers in CartPole-V0.

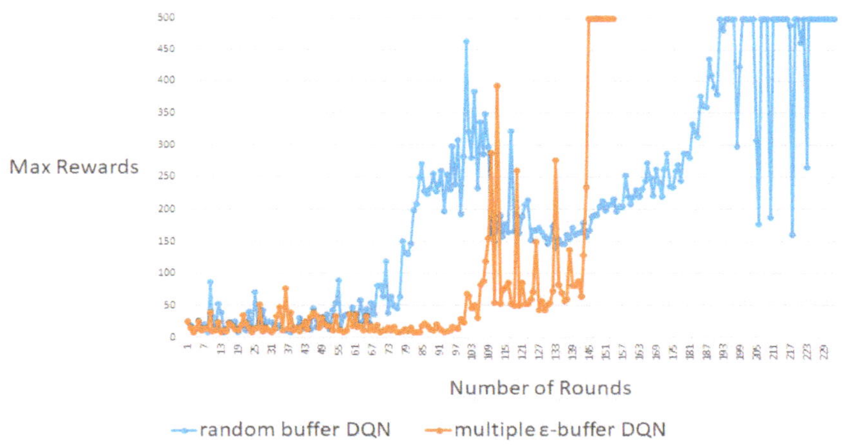

Figure 4. One of the best results between DQN with random buffers and DQN with multiple random ε-buffers in CartPole-V0.

4.2. *MountainCar-V0*

For DQN with multiple random ε-buffers, we follow similar previous studies [23]. Figure 5b shows a MountainCar-V0 [22] in OpenAI Gym [19]. Thus, GAMMA = 0.95, LEARNING − RATE = 0.001, MEMORY − SIZE = 100,000, BATCH − SIZE = 64, EXPLORATION − MAXIMUM = 1.0, EXPLORATION − MININUM = 0.01, and EXPLORATION − DECAY = 0.995 are the same as with previous DQN studies [23] because of a fairness comparison. In the MountainCar-V0 environment, there are two observations and three discrete actions, in Figure 6. Additionally, there are two episode terminations: when it reaches 0.5 position of the flag or if 200 iterations are reached. A penalty of 1 unit is applied for each move, including doing nothing. Our method limits the maximum number of times to 10,000 and 60 episodes by following the previous implementation of "Car has reached the goal" in each round [23].

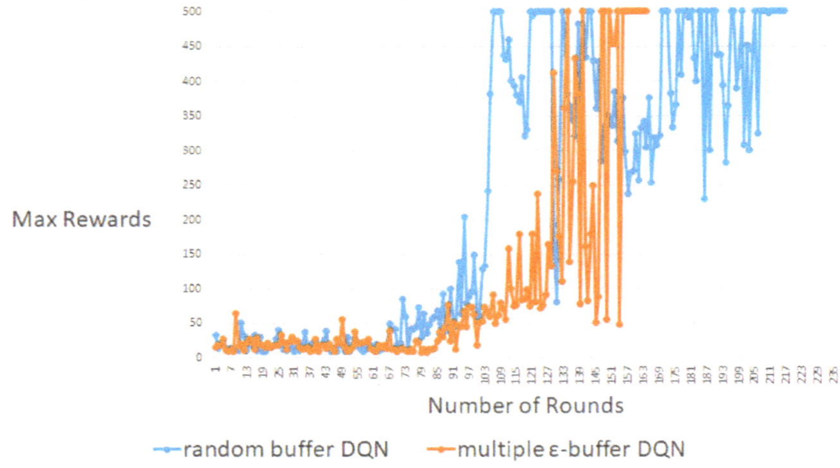

Figure 5. One of the worst results between DQN with random buffers and DQN with multiple random ε-buffers in CartPole-V0.

(a) Environment Observation

Number	Observation	Minimum	Maximum
0	Position	-1.2	+0.6
1	Velocity	-0.07	+0.07

(b) Actions

Number	Action
0	Push Left
1	No Push
2	Push Right

Figure 6. (**a**) The environment of MountainCar-V0 and (**b**) the actions taken by the agent of MountainCar-V0 [22].

Figures 7–9 display the maximum results for the number of rounds, from approximately 100 to 300. A DQN with multiple random ε-buffers can yield better results compared to when one random buffer is used. In each case, for around 100, 200 and 300 rounds, our proposed method shows that the number of "Car has reached the goal" [23] is higher than that obtained by DQN with one random buffer. We guarantee similar results for deep learning car simulations online, such as real-time learning with a normal neural network. Moreover, we can expect a real-time system with live training for real-time safeties. Similarly, for the experiment on Cart-Pole, we exploit the behavior policy to implement better exploration with multiple random ε-buffers so as not to follow a local minimum [9,10].

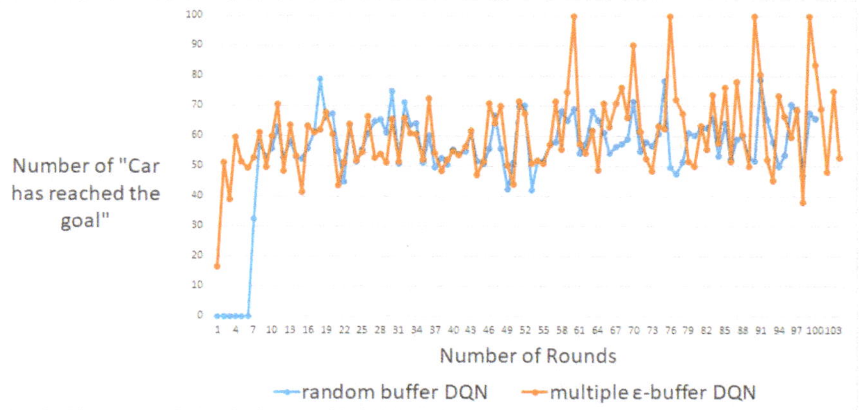

Figure 7. Maximum number of rounds, i.e., around 100, between DQN with random buffers and DQN with multiple random ε-buffers in MountainCar-V0.

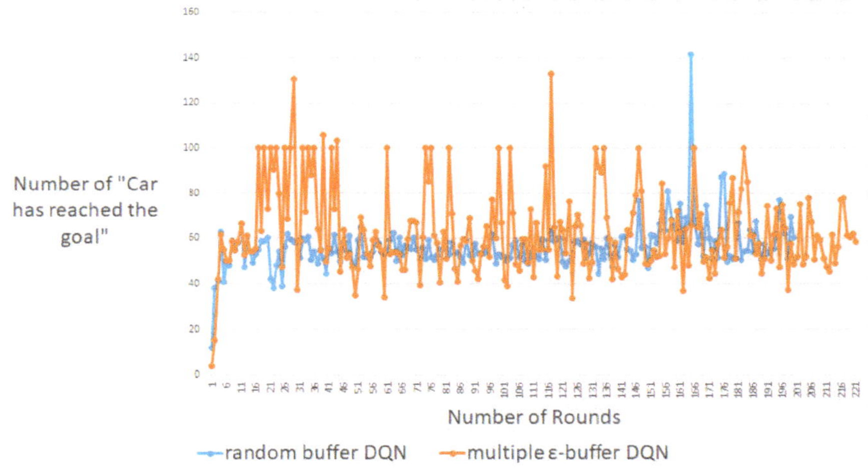

Figure 8. Maximum number of rounds, i.e., around 200, between DQN with random buffers and DQN with multiple random ε-buffers in MountainCar-V0.

4.3. Pendulum-V0

For DDPG with multiple random ε-buffers, we follow similar previous studies [25]. Figure 5c shows a Pendulum-V0 [24] by OpenAI Gym [19]. Thus, LEARNING – RATE = 0.001, MEMORY – SIZE = 2000, BATCH – SIZE = 32, EXPLORATION – MAXIMUM = 1.0, EXPLORATION – MINIMUM = 0.01, and EXPLORATION – DECAY = 1/2000 are the same as with the previous DQN studies [25] because of a fairness comparison. In the pendulum environment, there are three observations and one discrete action; see Figure 10. Additionally, there is a precise equation for the reward: $-(\theta^2 + 0.1 \times \theta^2 + 0.001 \times \text{action}^2)$, where θ is normalized between $-\pi$ and $+\pi$. Therefore, the lowest cost is $-(\pi^2 + 0.1 \times 8^2 + 0.001 \times 2^2) = -16.2736044$ and the highest cost is θ [25]. The goal is to remain at zero angle (vertical) with the least rotational velocity and least effort. There are no episode terminations, adding a maximum number of steps [24,25].

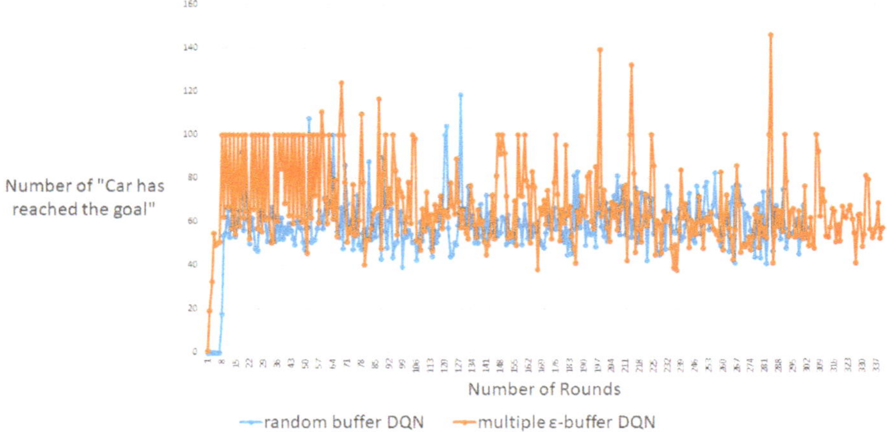

Figure 9. The maximum number of rounds, i.e., around 300, between DQN with random buffers and DQN with multiple random ε-buffers in MountainCar-V0.

(a) Environment Observation

Number	Observation	Minimum	Maximum
0	Cos(theta)	-1.0	+1.0
1	Sin(theta)	-1.0	+1.0
2	Theta dot	-0.8	+0.8

(b) Actions

Number	Action	Minimum	Maximum
0	Joint Effort	-2.0	+2.0

Figure 10. (**a**) The environment of MountainCar-V0 and (**b**) the actions taken by the agent of Pendulum-V0 [24].

Figures 11–13 exhibit the maximum results for the number of rounds, ranging from approximately 100–300. A DDPG with multiple random ε-buffers can yield a small improvement compared to that with one random buffer. In each case, for around 100, 200 and 300 rounds, our proposed method shows higher rewards than DDPG [16] with one random buffer. However, unlike the results of CartPole-V0 [20] and MountainCar-V0 [22], DDPG with multiple random ε-buffers cannot obtain a significant result compared with the cases of DQN. We consider this from the viewpoint of the on-policy. Fundamentally, a DDPG is based on the on-policy. Moreover, there is an issue about interpolating between policy optimization [2,16] and Q-learning [9,10]. The multiple random ε-buffers are actually based on Q-learning.

To train an agent, there are two major algorithms in model-free RL: policy optimization and Q-learning, which can follow the Bellman equation. A Q-learning agent selects an action to maximize the Q-value function based on the data at some point in the environment, regardless of the policy, known as the off-policy. Therefore, explorations on sample efficiency are extremely important to maximize the objective Q-value function. The representative algorithm in the off-policy is DQN. However, in terms of policy optimization, for better performance, an agent follows the recent deterministic policy and exploits a gradient ascent of the deterministic policy, thereby optimizing the parameters for the deterministic policy, known as the on-policy. Therefore, the update of the deterministic policy is extremely important for the exploration of the sample data. DDPG trains the agent by developing both the deterministic policy and the Q-value function. However, it is difficult to obtain both principled policy optimization and efficient Q-learning. That is why we cannot achieve better results in terms of DDPG. Therefore, we developed "Tradeoff Policy Optimization and Q-Learning" based on "Data vs. Policy" for better DDPG results. Thus, superior results can be obtained if we exploit a policy optimization method, based on the on-policy. This is another issue for future research.

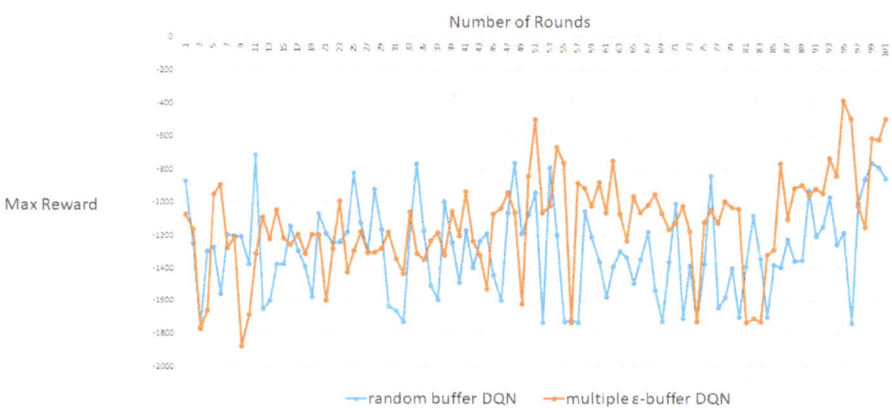

Figure 11. Maximum number of rounds, i.e., around 100, between DDPG with random buffers and DQN with multiple random ε-buffers in MountainCar-V0.

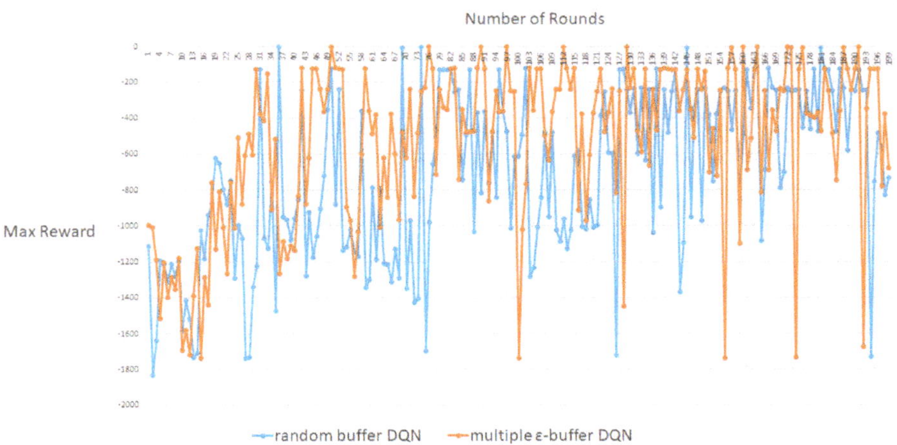

Figure 12. Maximum number of rounds, i.e., around 200, between DDPG with random buffers and DQN with multiple random ε-buffers in MountainCar-V0.

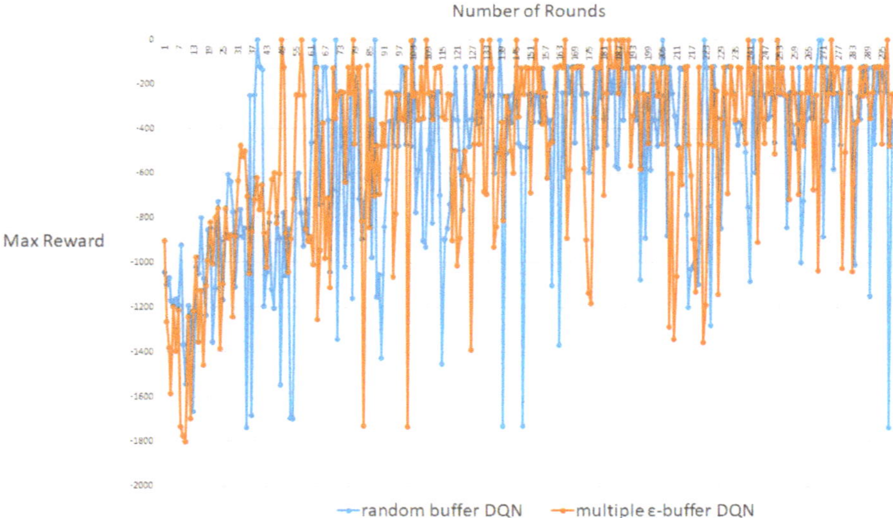

Figure 13. Maximum number of rounds, i.e., around 300, between DDPG with random buffers and DQN with multiple random ε-buffers in MountainCar-V0.

5. Conclusions

We proposed multiple ε-greedy experience buffers in off-policy deep RL to enhance the exploration for superior and near-perfect generalization. We considered strengthening the advantages of explorations of model-free deep RL. We exploited multiple random ε-buffers for one original goal, and emulated the environment of OpenAI Gym to achieve better results in environments such as CartPole-V0, MountCar-V0, and Pendulum-V0. Our results demonstrated that the off-policy method was beneficial through an experimental comparison of DQN and DDPG based on on-policy. The proposed model is compatible with discrete actions as well as continuous control, symmetrically. Therefore, we can expect better prediction accuracy in real-time online learning, as in the detection of network intrusions through the verification of whether the network is "normal or anomalous." Although our results are superior to those obtained by normal DQN with one random buffer, there is still scope for improvement, particularly for balancing policy optimization and Q-learning. The multiple random ε-buffers are fundamentally based on Q-learning. Therefore, the trade-offs between policy optimization and Q-learning should be solid and strong. Because of the deep neural network in RL, better results can be acquired in terms of DDPG, which is based on the on-policy of policy optimization. However, we can obtain better results if we develop another method based on policy optimization beyond the multiple random ε-buffers. The comparisons conducted in this study reveal that our model displays improvements in terms of exploration over the deep RL algorithms with Q-learning systems.

Author Contributions: Conceptualization, C.K. and J.P.; methodology, C.K.; writing—original draft preparation, C.K.; writing—review and editing, J.P.; project administration, J.P.; funding acquisition, J.P.

Funding: This research was supported by the Basic Science Research Program through the National Research Foundation of Korea (NRF) funded by the Ministry of Education (NRF2017R1D1A1B03035833).

Conflicts of Interest: The authors declare no conflict of interest.

References

1. Sutton, R.S.; Barto, A.G. *Reinforcement Learning: An Introduction*; MIT Press: Cambridge, MA, USA, 1998; Volume 1.

2. Haarnoja, T.; Zhou, A.; Abbeel, P.; Levine, S. Soft Actor-Critic: Off-Policy Maximum Entropy Deep Reinforcement Learning with a Stochastic Actor. *arXiv* **2018**, arXiv:1801.01290.
3. Kim, C.; Park, J. Designing online network intrusion detection using deep auto-encoder Q-learning. *Comput. Electr. Eng.* **2019**, *79*, 106460. [CrossRef]
4. Park, J.; Salim, M.M.; Jo, J.; Sicato, J.C.S.; Rathore, S.; Park, J.H. CIoT-Net: A scalable cognitive IoT based smart city network architecture. *Hum. Cent. Comput. Inf. Sci.* **2019**, *9*, 29. [CrossRef]
5. Sun, Y.; Tan, W. A trust-aware task allocation method using deep q-learning for uncertain mobile crowdsourcing. *Hum. Cent. Comput. Inf. Sci.* **2019**, *9*, 25. [CrossRef]
6. Kwon, B.-W.; Sharma, P.K.; Park, J.-H. CCTV-Based Multi-Factor Authentication System. *J. Inf. Process. Syst. JIPS* **2019**, *15*, 904–919. [CrossRef]
7. Srilakshmi, N.; Sangaiah, A.K. Selection of Machine Learning Techniques for Network Lifetime Parameters and Synchronization Issues in Wireless Networks. *Inf. Process. Syst. JIPS* **2019**, *15*, 833–852. [CrossRef]
8. Andrychowicz, M.; Wolski, F.; Ray, A.; Schneider, J.; Fong, R.; Welinder, P.; McGrew, B.; Tobin, J.; Abbeel, P.; Zaremba, W. Hindsight Experience Replay. *arXiv* **2017**, arXiv:1707.01495.
9. Mnih, V.; Kavukcuoglu, K.; Silver, D.; Graves, A.; Antonoglou, I.; Wierstra, D.; Riedmiller, M. Playing atari with deep reinforcement learning. *arXiv* **2013**, arXiv:1312.5602.
10. Silver, D.; Huang, A.; Maddison, C.J.; Guez, A.; Sifre, L.; Driessche, G.V.D.; Schrittwieser, J.; Antonoglou, I.; Panneershelvam, V.; Lanctot, M. Mastering the game of go with deep neural networks and tree search. *Nature* **2016**, *529*, 484–489. [CrossRef] [PubMed]
11. Cobbe, K.; Klimov, O.; Hesse, C.; Kim, T.; Schulman, J. Quantifying Generalization in Reinforcement Learning. *arXiv* **2019**, arXiv:1812.02341.
12. Liu, R.; Zou, J. The Effects of Memory Replay in Reinforcement Learning. *arXiv* **2017**, arXiv:1710.06574.
13. Schaul, T.; Quan, J.; Antonoglou, I.; Silver, D. Prioritized Experience Replay. *arXiv* **2016**, arXiv:1511.05952.
14. Plappert, M.; Houthooft, R.; Dhariwal, P.; Sidor, S.; Chen, R.; Chen, X.; Asfour, T.; Abbeel, P.; Andrychowicz, M. Parameter Space Noise for Exploration. *arXiv* **2018**, arXiv:1706.01905.
15. OpenReview.net. Available online: https://openreview.net/forum?id=ByBAl2eAZ (accessed on 16 February 2018).
16. Lillicrap, T.; Hunt, J.; Pritzel, A.; Heess, N.; Erez, T.; Tassa, Y.; Silver, D.; Wierstra, D. Continuous control with deep reinforcement learning. *arXiv* **2016**, arXiv:1509.02971.
17. FreeCodeCamp. Available online: https://medium.freecodecamp.org/improvements-in-deep-q-learning-dueling-double-dqn-prioritized-experience-replay-and-fixed-58b130cc5682 (accessed on 5 July 2018).
18. RL—DQN Deep Q-network. Available online: https://medium.com/@jonathan_hui/rl-dqn-deep-q-network-e207751f7ae4 (accessed on 17 July 2018).
19. OpenAI Gym. Available online: https://gym.openai.com (accessed on 28 May 2016).
20. Cart-Pole-V0. Available online: https://github.com/openai/gym/wiki/Cart-Pole-v0 (accessed on 24 June 2019).
21. Cart-Pole-DQN. Available online: https://github.com/rlcode/reinforcement-learning-kr/blob/master/2-cartpole/1-dqn/cartpole_dqn.py (accessed on 8 July 2017).
22. MountainCar-V0. Available online: https://github.com/openai/gym/wiki/MountainCar-v0 (accessed on 4 May 2019).
23. MountainCar-V0-DQN. Available online: https://github.com/shivaverma/OpenAIGym/blob/master/mountain-car/MountainCar-v0.py (accessed on 2 April 2019).
24. Pendulum-V0. Available online: https://github.com/openai/gym/wiki/Pendulum-v0 (accessed on 31 May 2019).
25. Pendulum-V0-DDPG. Available online: https://github.com/openai/gym/blob/master/gym/envs/classic_control/pendulum.py (accessed on 26 October 2019).
26. Tensorflow. Available online: https://github.com/tensorflow/tensorflow (accessed on 31 October 2019).
27. Keras Documentation. Available online: https://keras.io/ (accessed on 14 October 2019).

Article

SVD-Based Image Watermarking Using the Fast Walsh-Hadamard Transform, Key Mapping, and Coefficient Ordering for Ownership Protection

Tahmina Khanam [1], Pranab Kumar Dhar [1], Saki Kowsar [1] and Jong-Myon Kim [2,*]

[1] Department of Computer Science and Engineering, Chittagong University of Engineering and Technology (CUET), Chattogram-4349, Bangladesh; tahminacse0904079@gmail.com (T.K.); pranabdhar81@gmail.com (P.K.D.); sakikowsar@cuet.ac.bd (S.K.)

[2] School of IT Convergence, University of Ulsan, Ulsan 44610, Korea

* Correspondence: jmkim07@ulsan.ac.kr; Tel.: +82-52259-2217

Received: 27 October 2019; Accepted: 24 December 2019; Published: 26 December 2019

Abstract: Proof of ownership on multimedia data exposes users to significant threats due to a myriad of transmission channel attacks over distributed computing infrastructures. In order to address this problem, in this paper, an efficient blind symmetric image watermarking method using singular value decomposition (SVD) and the fast Walsh-Hadamard transform (FWHT) is proposed for ownership protection. Initially, Gaussian mapping is used to scramble the watermark image and secure the system against unauthorized detection. Then, FWHT with coefficient ordering is applied to the cover image. To make the embedding process robust and secure against severe attacks, two unique keys are generated from the singular values of the FWHT blocks of the cover image, which are kept by the owner only. Finally, the generated keys are used to extract the watermark and verify the ownership. The simulation result demonstrates that our proposed scheme is highly robust against numerous attacks. Furthermore, comparative analysis corroborates its superiority among other state-of-the-art methods. The NC of the proposed method is numerically one, and the PSNR resides from 49.78 to 52.64. In contrast, the NC of the state-of-the-art methods varies from 0.7991 to 0.9999, while the PSNR exists in the range between 39.4428 and 54.2599.

Keywords: fast Walsh–Hadamard transform; Gaussian mapping; singular value decomposition; coefficient ordering; key mapping

---

1. Introduction

The flow of multimedia data increases manifold with the recent infrastructural development of computer networks. Accordingly, the proof of ownership issue for multimedia data has come to the surface as an impending challenge. In a bid to negotiate with this problem, the watermarking approach might be used as an indispensable tool. Since multimedia data often suffer from different types of transmission channel attacks, the technique should be immune to such maladies. Hence, the watermarking approach is used for hiding the digital information during transmission. The watermark is typically used to prove the ownership of such host signals. Several algorithms have been proposed in the literature to create robust and imperceptible watermarks. In general, watermarking methods can be divided into three main categories: (i) blind methods [1–13], (ii) semi-blind methods [14–17] (iii) non-blind methods [18–24]. A blind watermarking framework for high dynamic range images (HDRIs) is proposed in [1]. In this method, the artificial bee colony algorithm is employed to select the best block for the embedding watermark. Then, the watermark is inserted in the first level approximation sub-band of the discrete wavelet transform (DWT) of each selected block. This method provides good quality watermarked images, although it is not robust against geometric attacks such as rotation and scaling. In [2], a new blind error diffusion-based halftone visual watermarking method called content aware double-sided embedding error diffusion (CaDEED) is introduced. By adopting the problem formulation of CaDEED, the optimization problem is solved in order to achieve an optimal solution. Although it shows good results for imperceptibility and robustness, the performance of this system is highly dependent on the content of the host image and watermark. A blind integer wavelet-based watermarking scheme for inserting the compressed version of the binary watermark is presented in [3]. The peak signal-to-noise ratio (PSNR) result of this method is quite satisfactory. However, the robustness against compression attacks is not significant. The authors in [4] proposed a blind geometrically invariant image watermarking method by employing connected objects and a gravity center. This framework has proven resistant against geometrical attacks, such as rotation and scaling. However, it has low robustness against other regular noise attacks such as Gaussian or speckle noise. Furthermore, a contrast-adaptive strategy as a removal solution for visible watermarks is presented in [5] where a sub-sampling technique is adopted to propose such a blind system. The imperceptibility results of this method are very good. However, it shows low robustness against some attacks. In addition to this, a blind watermarking scheme based on singular value decomposition (SVD) is introduced in [6]. Initially, they analyzed the orthogonal matrix U via SVD. This work utilizes the concept of finding a strong similarity correlation existing between the second-row first column element and the third-row first column element. At the final stage, the color watermark is embedded by slightly modifying the value of the second-row first column element and the third-row first column element of the U matrix. The technique performs well against various attacks, although it demonstrates very poor performance under median filtering of the watermarked image. Furthermore, the authors in [7] proposed a robust watermarking scheme using discrete cosine transform (DCT) and SVD for lossless copyright protection. Its imperceptibility result is significantly good. However, its robustness result against cropping attacks is quite low. A blind simple watermarking algorithm for image authentication is presented using fractional wavelet packet transform (FRWPT) and SVD in [8]. The proposed algorithm performs the embedding operation on singular values of the host image. To improve the fidelity, the perceptual quality of the watermarked images is exhibited. Although this method is highly secured, it shows low robustness against various attacks for some watermarked images. For estimation of the original coefficients, a blind watermarking method is placed in [9]; the authors used a trained SVR there. Additionally, the particle swarm optimization (PSO) is further utilized to optimize the proposed scheme. It provides high imperceptibility; however, it could not show excellent robustness against several attacks. A blind self-synchronized watermarking method in the cepstrum domain is suggested in [10]. This method does not provide a good trade-off between imperceptibility and robustness. Furthermore, a blind scheme is proposed in [11] in a bid to obtain minimal image distortion. This method provides high-quality watermarked images, albeit low robustness against various attacks. In [12], hamming codes are used to embed the authentication information in a cover image. The watermark extraction process of this method is blind and provides satisfactory results in imperceptibility. However, the robustness result against various attacks is not reported there. The authors of [13] suggested a blind watermarking algorithm based on lower-upper (LU) decomposition. The watermark is embedded into the first-column second-row element and the first-column third-row element of the lower triangular matrix obtained from LU decomposition. It provides good quality watermarked images despite the low robustness against compression attacks. A semi-blind self-reference image watermarking method using discrete cosine transform (DCT) and singular value decomposition (SVD) is proposed in [14]. Initially, essential blocks are fetched by using a threshold on the number of edges in each block. Using these essential blocks, a reference image is created and then transformed into the DCT and SVD domain. Embedding the watermark is done by modifying singular values of the host image using singular values of the watermark image. This method yields good quality watermarked images. However, it shows low robustness against the scaling operation. To embed the watermark, the concepts of vector quantization (VQ) and association rules in data mining are employed in [15]. The approach is semi-blind, which hides the association rules of the watermark instead of the whole watermark. This method shows good robustness against various attacks with poor performance on imperceptibility. In addition, a reference watermarking scheme with semi-blind is proposed in [16] based on DWT and SVD for copyright protection and authenticity. The method has high imperceptibility showing the low robustness against cropping and rotation attacks. An image watermarking method using DWT, all phase discrete cosine bi-orthogonal transform (APDCBT), and SVD is proposed in [17]. This method shows high imperceptibility; however, it provides low robustness against combined cropping and compression attacks. A non-blind image watermarking algorithm based on the Hadamard transform is proposed in [18]. In this method, the breadth first search (BFS) technique is used to embed the watermark. Notably, it shows good performance in imperceptibility. However, it has the limitation of relatively poor performance against compression attacks. The authors in [19] introduced a non-blind robust watermarking technique using DCT and a normalization procedure. They used image normalization for calculating the affine transform parameters so that the watermark embedding and detection processes can be performed in the original coordinate system. However, this method shows low robustness against some attacks. In [20], a non-blind digital watermarking algorithm using wavelet-based contourlet transform (WBCT) is presented. To select the position for inserting the watermark, the texture information of the image is used. It has good robustness against numerous attacks, albeit low robustness against filtering attacks. Moreover, the imperceptibility result of this method is not reported there. A non-blind hybrid image watermarking scheme based on DWT and SVD is proposed in [21]. In this approach, the watermark is embedded to the elements of singular values of the cover image of DWT sub-bands. The imperceptibility result of this method is quite high, having low the robustness against cropping attacks. A non-blind SVD-based digital watermarking scheme for ownership protection is proposed in [22]. In this method, a meaningful text message is used rather than using a randomly generated Gaussian sequence. However, the robustness of this method against attacks is low. A non-blind image watermarking using DCT and DWT is proposed in [23]. The DCT coefficients of the watermark image are embedded into four DWT bands of the color components of the host image. The imperceptibility result of this method is quite satisfactory. However, the robustness against rotation attacks is a little low. A non-blind color image watermarking method using SVD and QR code is suggested in [24]. This method shows good results in imperceptibility; the robustness result against poison and speckle noise attack is not reported.

From the above studies, we can conclude that some methods have low robustness, whereas some methods have less imperceptible or less secured. Further, some methods are non-blind and semi-blind. To overcome these limitations, an SVD-based blind symmetric image watermarking method using fast Walsh–Hadamard transform (FWHT) with key mapping and coefficient ordering for ownership protection is proposed in this paper. In symmetric watermarking, the same keys are used for embedding and detecting the watermark. The major contributions of this research work are subjected:

- A blind image watermarking method is proposed that is highly robust and secured against numerous attacks while providing good quality watermarked images;
- To safeguard the unauthorized detection, the Gaussian mapping is used to scramble the watermark;
- To facilitate authentic and errorless extraction of the watermark image by generating the keys from the singular values the FWHT blocks of the cover image;
- It provides a good trade-off among robustness, security, and imperceptibility.

Simulation results indicated that our proposed method is highly robust against numerous attacks. The normalized correlation (NC) of the proposed method is numerically one, whereas the NC of the recent methods [13,23,24] vary from 0.7991 to 0.9999. The peak signal-to-noise ratio (PSNR) of the proposed method varies from 49.78 to 52.64, whereas the PSNR of the recent methods [13,23,24] vary from 39.4428 to 54.2599. In other words, the proposed method outperforms state-of-the-art methods in terms of robustness, security, and imperceptibility.

The rest of the paper is organized as follows. Section 2 introduces the background information, whereas the proposed watermarking method is illustrated in Section 3. Section 4 provides the experimental results. Finally, the paper is concluded in Section 5 with future remarks.

2. Background Information

2.1. Singular Value Decomposition

For an $M \times M$ square matrix X with rank $\leq M$, its SVD is represented by Equation (1):

$$X = UDV^T$$

$$X = \begin{bmatrix} u_{1,1} & \cdots & u_{1,M} \\ u_{2,1} & \cdots & u_{2,M} \\ \vdots & \ddots & \vdots \\ u_{M,1} & \cdots & u_{M,M} \end{bmatrix} \begin{bmatrix} \lambda_1 & 0 & \cdots & 0 \\ 0 & \lambda_2 & \cdots & 0 \\ \vdots & \vdots & \ddots & \vdots \\ 0 & 0 & \cdots & \lambda_M \end{bmatrix} \begin{bmatrix} v_{1,1} & \cdots & v_{1,M} \\ v_{2,1} & \cdots & v_{2,M} \\ \vdots & \ddots & \vdots \\ v_{M,1} & \cdots & v_{M,M} \end{bmatrix} \quad (1)$$

where U and V are $M \times M$ orthogonal matrices, and D is a singular diagonal matrix with diagonal elements $\lambda_1, \lambda_2, \lambda_3, \ldots, \lambda_M$,. These diagonal elements are unique for image data. Therefore, these values are used to generate unique keys for the errorless and authentic extraction of the watermarks.

2.2. Fast Walsh-Hadamard Transform

General Hadamard transform is performed by a Hadamard matrix H with the size 4×4 defined in Equation (2). It is an orthogonal square matrix with only +1 and −1 values. Furthermore, it has a unique sequence that is counted on the basis of the changes of the values in a row.

$$H = \begin{bmatrix} 1 & 1 & 1 & 1 \\ 1 & -1 & 1 & -1 \\ 1 & 1 & -1 & -1 \\ 1 & -1 & -1 & 1 \end{bmatrix} \quad (2)$$

The Hadamard transform concentrates most of the energy into the upper left corner of the transformed matrix. The direct current (DC) and alternating current (AC) coefficients of the transform matrix are arranged in zigzag order from low-frequency components to high-frequency components. In this study, the low-frequency components are used for embedding the watermark, since they are less sensitive to noise. Additionally, the Hadamard matrix has a different form called Walsh matrix W, which is defined in Equation (3).

$$W = \begin{bmatrix} 1 & 1 & 1 & 1 \\ 1 & 1 & -1 & -1 \\ 1 & -1 & -1 & 1 \\ 1 & -1 & 1 & -1 \end{bmatrix} \quad (3)$$

In this proposed method, fast Walsh–Hadamard transform (FWHT) is utilized, which is a technique of calculating a discrete Walsh–Hadamard transform with less computation time.

3. Proposed Method

Let $X = \{x(i,j), 1 \leq i \leq M,\ 1 \leq j \leq M\}$ be the original host image and $W = \{w(k,l), 1 \leq k \leq N,\ 1 \leq l \leq N\}$ be the watermark image to be embedded into the original image.

3.1. Watermark Preprocessing

It is essential to preprocess the watermark for enhancing its security. Preprocessing includes the scrambling of the watermark image. In this proposed method, we utilize Gaussian mapping to scramble the watermark. To implement the Gaussian mapping on the watermark, the following steps are performed:

Step 1. The watermark image W is reshaped into a one-dimensional sequence $Q = \{q(r), 1 \leq r \leq N \times N\}$.

Step 2. Initially, a reference pattern $P = \{p(r),\ 1 \leq r \leq N \times N\}$ is generated using a Gaussian map, which is defined in Equation (4).

$$p(r) = \exp\left\{-\mathrm{a}\times(p(r+1))^2\right\}+\mathrm{b} \quad (4)$$

where a, b, and $p(1)$ are predefined constants and are used as key $k3$, as shown in Figure 1.

Step 3. Then, the binary reference pattern $Z = \{z(r),\ 1 \leq r \leq N \times N\}$ is calculated using the following equation:

$$z(r) = \begin{cases} 1 & if\ p(r) > T \\ 0 & otherwise \end{cases} \quad (5)$$

where T is a predefined threshold.

Step 4. Finally, the watermark sequence $q(r)$ is scrambled with $z(r)$ using Equation (6):

$$u(r) = z(r) \oplus q(r),\quad 1 \leq r \leq N \times N \quad (6)$$

where $\oplus$ denotes the bitwise XOR operation.

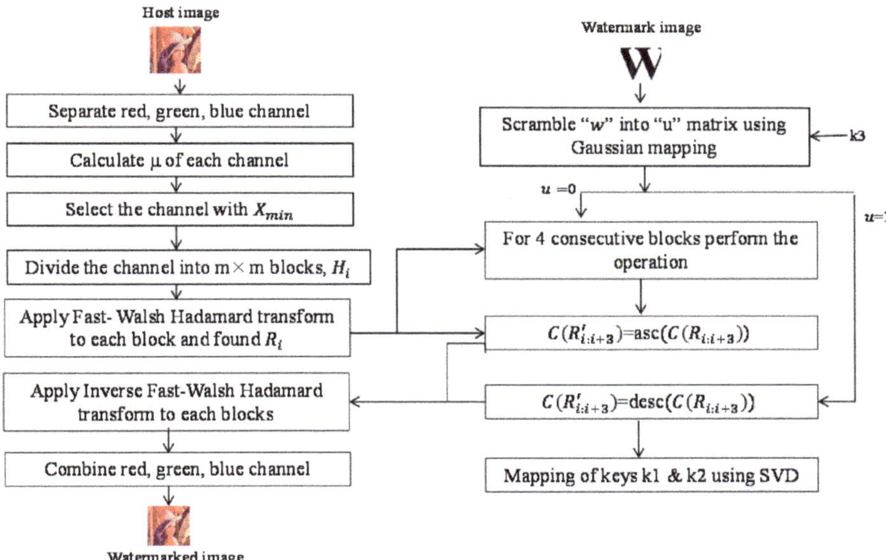

Figure 1. Proposed embedding algorithm.

3.2. Watermark Embedding Process

The proposed watermark embedding process is shown in Figure 1. The pseudo code of the watermark embedding process is presented in Algorithm 1. The embedding process is described in the following steps:

Step 1. The original host image X is first divided into three channels X_{red}, X_{green}, and X_{blue} , where X_{red}, X_{green}, and X_{blue} represent the red, green, and blue channels of the original image, respectively. Then, the mean of the pixel values of each channel is calculated using Equation (7).

$$\mu(X_{red}) = \sum_{i=1}^{M}\sum_{j=1}^{M}\frac{X_{red}}{255},\ \mu(X_{green}) = \sum_{i=1}^{M}\sum_{j=1}^{M}\frac{X_{green}}{255},\ \mu(X_{blue}) = \sum_{i=1}^{M}\sum_{j=1}^{M}\frac{X_{blue}}{255} \tag{7}$$

where $\mu(X_{red})$, $\mu(X_{green})$, and $\mu(X_{blue})$, indicate the mean of the pixel values of the red, green, and blue channels, respectively. After that, the channel with minimum mean X_{min} is selected, which is either X_{red}, X_{green}, or X_{blue}.

Step 2. The selected channel X_{min} is further divided into $m \times m$ non-overlapping blocks, $H = \{H_i;\ 1 \leq i \leq n\}$, where i is the block number and m is the length of the row and column of each block.

Step 3. FWHT is applied in each block H_i to obtain the transformed block R_i, where R_i contains the FWHT coefficients.

Step 4. Among all the n blocks, each set of four consecutive blocks R_i, R_{i+1}, R_{i+2}, and R_{i+3} is selected to embed a watermark bit. The main idea of the embedding process is to sort the coefficients of the first row represented by $C\left(R_{i:i+3}\right)$, where $\{i : i+3\}$ indicates $\{i, i+1, i+2, i+3\}$ of each set of selected blocks R_i, R_{i+1}, R_{i+2}, and R_{i+3} except the DC value. If the watermark bit is 1, the selected low-frequency coefficients $C\left(R_{i:i+3}\right)$ are sorted in descending order; otherwise, they are sorted in ascending order. The concept of embedding the watermark bit in ascending and descending order with a block size of 4×4 where, $m = 4$, is shown in Figure 2.

120	1	2	4
-	-	-	-
-	-	-	-
-	-	-	-

(**a**)

120	4	2	1
-	-	-	-
-	-	-	-
-	-	-	-

(**b**)

Figure 2. (**a**) Sorting in ascending order to embed 0 bits and (**b**) sorting in descending order to embed 1 bit.

In this step, two keys $k1$ and $k2$ are also used in order to make the watermarking method more secured. The key $k1$ is generated from the singular values of each block H_i of the selected channel of host image. The key $k2$ is generated from key $k1$, which is used to authenticate key $k1$ in the watermark extraction process. The following operation is performed for embedding a watermark bit into each selected block.

$$\begin{array}{ll} C\left(R'_{i:i+3}\right) = asc\left(C\left(R_{i:i+3}\right)\right);\ mapping\ key\ k1\ and\ k2, & when\ u(r) = 0 \\ C\left(R'_{i:i+3}\right) = desc\left(C\left(R_{i:i+3}\right)\right);\ mapping\ key\ k1\ and\ k2, & when\ u(r) = 1 \end{array} \tag{8}$$

where *asc* and *desc* represent sorting the data in ascending order and descending order, respectively. The process of mapping the keys $k1$ and $k2$ are described in the next section.

Mapping key $k1$ and $k2$: In this section, the process of mapping keys $k1$ and $k2$ is explained, which is defined in Equation (8). This step is introduced to strengthen the proposed algorithm under severe attack. To map the keys initially, SVD is applied in each block H_i to generate the necessary information. To perform the operation, the following steps are used:

(1) Each block H_i of the selected channel is decomposed into three matrices: U_i, D_i, and V_i using Equation (9).

$$H_i = U_i D_i V_i^T \tag{9}$$

where λ_{i1}, λ_{i2}, ..., λ_{im} are the singular values of the matrix D_i of each block H_i. These singular values are unique for each block H_i. The keys $k1$ and $k2$ are calculated using these singular values. Thus, unauthorized people could not map the keys without the host image to prove fake ownership. To do this, initially, a null key $k1$ is defined. Then, $k1$ is generated using these singular values as defined in Equation (10) below:

$$k1 = append\Big(k1,\ \Big(asc\Big(\lambda_{ij}\Big)\Big)\Big),\ u(r) = 0k1 = append\Big(k1,\ \Big(desc\Big(\lambda_{ij}\Big)\Big)\Big),\ u(r) = 1 \tag{10}$$

where i indicates the block number, $j = \{1 \le j \le m\}$ indicates the singular values of each block, *asc* and *desc* represent sorting the data in ascending order and descending order, respectively, and *append* indicates the concatenation operation. The singular values are sorted according to the watermark bit 0 or 1.

(2) Finally, $k1$ is converted into a one-dimensional sequence of length $L = n \times m$, where n is the total number of blocks and m is the total number of singular values in each block.

(3) To generate key $k2$, define a null key $k2$ with length S, where $S = n/m$. Then, $k2$ is generated from key $k1$ using the following Equation (11):

$$k2 = append(k2, \mu\ (k1_{h:h+t}))\ where\ 1 \le h \le L \tag{11}$$

where μ is the mean of consecutive t values of key $k1$ and the length of key $k2$ is $(n/m) + 1$. Although the first key can be generated by the owner only, the second key is generated to authenticate the first key in the extraction process.

Step 5. Inverse FWHT is applied to each transformed block R'_i and the watermarked blocks H'_i are found.

Step 6. Finally, three watermarked channels $X'_{red}, X'_{green},$ and X'_{blue} are combined to generate the watermarked image X'.

Algorithm 1: Watermark Insertion

Variable Declaration:
X: Host image
μ: Mean intensity value of each channel of host image (Lena)
X_{min}: Channel with minimum mean
H_i: Non-overlapping blocks of X_{min} (size 4 × 4)
FWHT, SVD: Transformation and decomposition used in the algorithm
R_i: FWHT transformed block of H_i
$C(R_{i:i+3})$: Three coefficients of first row (except DC value) of the consecutive transformed block
$C(R'_{i:i+3})$: Coefficients in ascending or descending order
W: Watermark image
u: Scrambled watermark sequence
Watermark Embedding Procedure:
1. Watermark preprocess: scramble W to obtain u using Gaussian mapping
2. Read the host image and calculate μ of each channel (Red, Green, Blue)
X.bmp (host image with size of 256 × 256)
W.bmp (watermark image with size of 32 × 32)
3. Select channel X_{min} and divide it into 4 × 4 H_i blocks
4. Apply FWHT to each block H_i and found R_i
5. Watermark Insertion

$$C\Big(R'_{i:i+3}\Big) = asc\Big(C\Big(R_{i:i+3}\Big)\Big);\ mapping\ key\ k1\ and\ k2,\ when\ u(r) = 0$$

$$C\Big(R'_{i:i+3}\Big) = desc\Big(C\Big(R_{i:i+3}\Big)\Big);\ mapping\ key\ k1\ and\ k2,\ when\ u(r) = 1$$

asc: ascending order, desc: descending order, $1 \le r \le 32 \times 32$
// Use SVD to map keys $k1$ and $k2$
6. Perform inverse FWHT and combine the channels to get the Watermarked Image

---

3.3. Watermark Detection Process

The watermark extraction process has two main phases: (1) modify the degree of ascendant/descendant of the attacked watermarked image with key $k1$, and (2) authenticate key $k1$ with key $k2$. In addition, the pseudo code of the watermark detection process is presented in Algorithm 2. The overall process is described below and shown in Figure 3:

Step 1. The attacked watermarked image X^* is first divided into three channels, $\{X^*_{red}, X^*_{green}, and\ X^*_{blue}\}$. Then, the mean value of the pixels of the red, green, and blue channels represented by $\mu(X^*_{red})$, $\mu(X^*_{green})$, and $\mu(X^*_{blue})$ are calculated. Thereafter that, the channel with minimum mean X^*_{min} is selected for extracting the watermark.

Step 2. The selected channel X^*_{min} is further divided into $m \times m$ non-overlapping blocks H^*_i, where i is the block number.

Step 3. FWHT is carried out on each block H^*_i. After applying this operation, the transformed blocks R^*_i are found.

Step 4. The degree of ascendant/descendant denoted by *dof* is calculated for four consecutive transformed blocks $\{R^*_i, R^*_{i+1}, R^*_{i+2}, R^*_{i+3}\}$. Therefore, *dof*(*asc*) represents the number of times that the low-frequency coefficients in the first row $C^*\left(R^*_{i:i+3}\right)$ of each transformed block except for the DC value are in ascending order. Similarly, the *dof*(*desc*) represents the number of times that the low-frequency coefficients in the first row $C^*\left(R^*_{i:i+3}\right)$ of each transformed block except the DC value are in descending order.

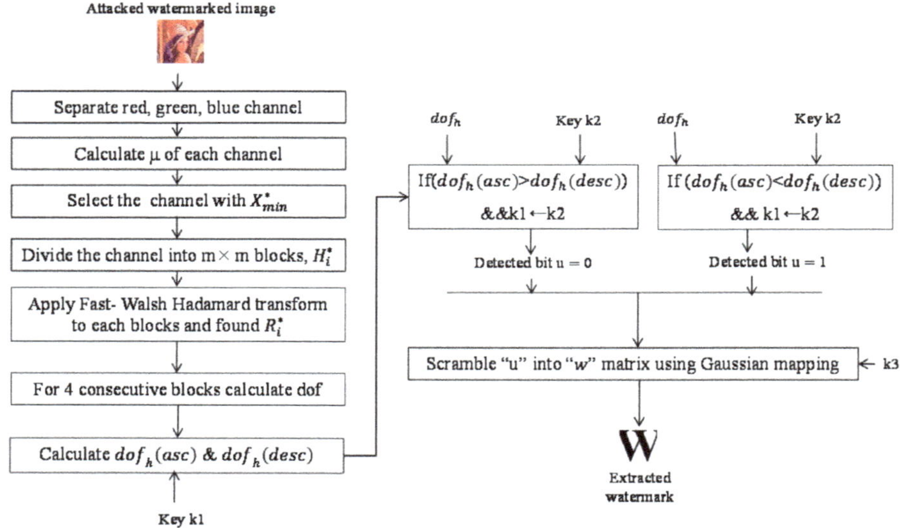

Figure 3. Proposed extraction algorithm.

Later, dof is modified with key $k1$. This phase assists the system to resist when the noise attack is severe. Initially, the dof' of the first t values of key $k1$ is calculated to extract the first watermark bit using Equations (12) and (13). Thus, consecutive t values of the key are considered each time for extracting a one-bit watermark. We found another two matrices, $dof'(asc)$ and $dof'(desc)$, with $\frac{L}{t} = N^2$ values, where L is the length of key $k1$, with $1 \leq h \leq L$.

$$dof'(asc) = dof'(asc) + 1;\ if\ k1_h > k1_{h+1} \tag{12}$$

$$dof'(desc) = dof'(desc) + 1;\ fk1_h < k1_{h+1} \quad (13)$$

Finally, we modify the dof with dof' by a simple addition operation, as shown in Equations (14) and (15), and found two matrices, $dof_h(asc)$ and $dof_h(desc)$.

$$dof_h(asc) = dof(asc) + dof'(asc) \quad (14)$$

$$dof_h(desc) = dof(desc) + dof'(desc) \quad (15)$$

Authenticate $k1$ with $k2$: This operation is carried out to authenticate key $k1$ using $k2$. For this purpose, the average of the consecutive t values of $k1$ is calculated and compared with one value of $k2$. This operation is represented using Equations (16) and (17) given below:

$$if(\mu(k1_{h:h+t})) = k2_h;\ k1 \leftarrow k2 \quad (16)$$

$$if(\mu(k1_{h:h+t}))! = k2_h;\ !k1 \leftarrow k2 \quad (17)$$

where $k1 \leftarrow k2$ means $k1$ is authenticated by $k2$ and $!k1 \leftarrow k2$ means $k1$ is not authenticated by $k2$. If $k1$ is authenticated, then the watermark would be extracted accordingly.

Step 5. The hidden binary sequence is found using the following rule:

If $dof_h(asc) > dof_h(desc)$ *and* $k1 \leftarrow k2$
then $u(r) = 0$
else If $dof_h(asc) > dof_h(desc)$ *and* $k1 \leftarrow k2$
then $u(r) = 1$

Step 6. The binary watermark sequence $q^*(r)$ is extracted with key $k3$ using the following equation:

$$q^*(r) = z(r) \oplus u(r), \quad 1 \le r \le N \times N. \quad (18)$$

Finally, the watermark image W^* is found by arranging the watermark sequence $q^*(r)$ into the $N{\times}N$ matrix.

Algorithm 2: Watermark Extraction

Variable Declaration:
X^*: Attacked watermarked image
μ: Mean intensity value of each channel of X^*
X^*_{min}: Channel with minimum mean
H^*_i: Non-overlapping blocks of X^*_{min} (size 4×4)
FWHT: Transformations used in the algorithm
R^*_i: FWHT transformed block of H^*_i
$C^*(R^*_{i:i+3})$: Three coefficients of first row (except DC value) of four consecutive transformed block
W: Watermark image
u: Scrambled watermark sequence
dof(asc/desc): The number of times low-frequency coefficients in the first row $C^*(R^*_{i:i+3})$ of each transformed block except the DC value are in ascending/descending order.
Watermark Extraction Procedure:
1. Read X^* and calculate μ of each channel (Red, Green, Blue)
2. Select channel X^*_{min} and divide into 4×4 H^*_i blocks
3. Apply FWHT to each block H^*_i and found R^*_i
4. Watermark extraction

(a) Modifying *dof*(*asc*/*desc*) into *dof′*(*asc*/*desc*) with key $k1$

$$dof'(asc) = dof'(asc) + 1;\ if\ k1_h > k1_{h+1}$$

$$dof'(desc) = dof'(desc) + 1;\ f\ k1_h < k1_{h+1}$$

where L is the length of key $k1$, *with* $1 \le h \le L$ and then calculate

$$dof_h(asc) = dof(asc) + dof'(asc)$$

$$dof_h(desc) = dof(desc) + dof'(desc)$$

(b) Authenticate key $k1$ with key $k2$

$$if(\mu(k1_{h:h+t})) = k2_h;\ k1 \leftarrow k2$$

$$if(\mu(k1_{h:h+t}))! = k2_h;\ !k1 \leftarrow k2$$

where $k1 \leftarrow k2$ means $k1$ is authenticated by $k2$ and $!k1 \leftarrow k2$ means $k1$ is not authenticated by $k2$.
// Consecutive t values of the key are considered each time for extracting a one-bit watermark, where $\frac{L}{t} = N^2$ and $\mu(k1_{h:h+t})$ means mean of these t values
(c) Watermark extraction
If $dof_h(asc) > dof_h(desc)$ *and* $k1 \leftarrow k2$
then $u(r) = 0$
else If $dof_h(asc) > dof_h(desc)$ *and* $k1 \leftarrow k2$
then $u(r) = 1$
where, $1 \le r \le 32 \times 32$
(d) Re-scramble u to get W

4. Experimental Results and Discussions

In this section, the performance of our proposed method is evaluated in terms of imperceptibility and robustness. The proposed method used various images, including Lena, Peppers, Baboon, and Fruit, with the size 256 × 256 as host images shown in Figure 4. The size of the binary watermark image is 32 × 32, as shown in Figure 5. It performs well for all the host images in term of imperceptibility and robustness. In this study, the selected values for m and t are 4 and 16, respectively, as the size of each block H_i is 4 × 4. Therefore, the total number blocks is 4096. Thus, the length of the key *k1* is 16384. The main reason for selecting a smaller value for m to embed the watermark bit is that sorting larger blocks causes greater degradation in the quality of the watermarked image.

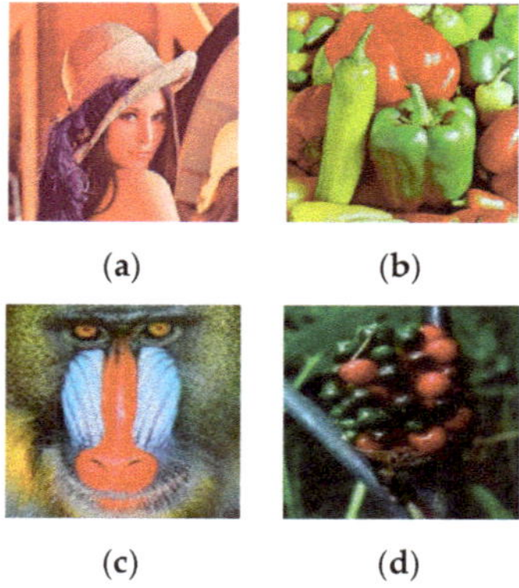

Figure 4. The host images: (**a**) Lena, (**b**) Peppers, (**c**) Baboon, and (**d**) Fruit.

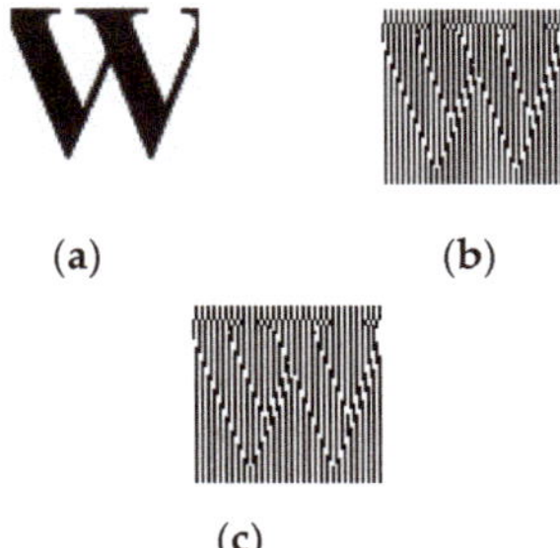

Figure 5. Watermark images: (**a**) original, (**b**) scrambled with a = 10, b = 0.05, and y0 = 20, and (**c**) scrambled with a = 30, b = 0.01, and y0 = 10.

Imperceptibility test: The imperceptibility of the watermarked images can be evaluated in terms of the peak signal-to-noise ratio (PSNR), as given in Equation (19).

$$\mathrm{PSNR} = 10\log_{10}\left(\frac{255^2}{\frac{1}{MM}\sum_{i=1}^{M}\sum_{j=1}^{M}(X - X')^2}\right) \tag{19}$$

where X and X' are the original and watermarked images, respectively. Higher values of PSNR indicate the better quality of the watermarked image. Figure 4 shows the original and scrambled images with different values of a, b, and y0.

To test the imperceptibility of the proposed framework, the PSNR values are calculated and compared with those values of the existing methods, as shown in Table 1. From this table, it is observed that the PSNR of the proposed method varies from 49.78 to 52.64, whereas the PSNR of the recent methods [13,23,24] varies from 47.1961 to 47.1836, 54.2577 to 54.2599, and 39.4428 to 40.8216. Therefore, it is evident that the PSNRs of the recent methods [23,24] are quite high, whereas the PSNRs of the recent method [13] are low compared to all other methods. In other word, the PSNR of the proposed method is higher than that of the methods reported in [13]. However, it is slightly lower than that of the method reported in [24]. This comparison justifies that the suggested method outperforms the other recent techniques. Since in each 4 × 4 block, only three AC values are shuffled, and the DC value remains in its position, low image degradation took place. However, low image degradation results in high imperceptibility.

Security analysis: For a secured watermarking method, how it performs against various attacks is very important. The proposed method utilizes a Gaussian map to enhance the security. To encrypt the watermark image, some predefined constants are used such as a, b, and $p(1)$, which are considered as secret key k3. If the selected value for a, b, and $p(1)$ are wrong, in that case, the watermark will not be extracted properly. Further, in order to make the watermarking method more secured, the two keys $k1$ and $k2$ are used. The key $k1$ is generated from the singular values of each block H_i of the selected channel of host image. Moreover, it is observed that these singular values are floating point numbers, and it is not possible to find out these singular values without the host image. Therefore, it is not possible to generate key $k1$ without the host image. The key $k2$ is generated from key $k1$, which is used to authenticate the key $k1$ in the watermark extraction process. Therefore, it is not possible to generate the key $k2$ without $k1$. These keys ($k1$, $k2$, $k3$) are used in the watermark detection process to extract the embedded watermark. The correct watermark can be extracted when all the keys ($k1$, $k2$, and $k3$) are correct. In other words, if any one of the keys is wrong, then the watermark will not be extracted correctly. This phenomenon is illustrated in Figure 6. Moreover, the size of each block H_i of the selected channel of the host image is 4 × 4; therefore, the total number of blocks in each host image is 4096. Thus, the length of the key $k1$ is 4096 × 4 = 16,384, and the length of the key $k2$ is (4096/4) + 1 = 1025, which are quite long, indicating that the key space is large enough. As the key $k1$, $k2$, and $k3$ are floating point numbers, therefore, the value of these keys cannot be determined. Hence, the probability of extracting the right watermark is near to 0. Therefore, the attacker cannot detect the correct watermark without the right key, which enhances the security of the proposed watermarking method.

Table 1. Comparison between the proposed and recent methods in terms of peak signal-to-noise ratio (PSNR).

Watermarked Images	Proposed Method	Ahmed et al. [23]	Patvardhar et al. [24]	Su et al. [13]
	50.04		54.2577	39.4428
	49.78	47.1961		40.8216
	51.56	47.1836	54.3499	
	52.64			

Key	Case 1	Case 2	Case 3	Case 4
Key $k1$	√	×	√	√
Key $k2$	√	√	×	√
Key $k3$	√	√	√	×
Extracted watermark	 W			

Figure 6. The extracted watermark with right and different wrong keys.

Robustness test: To measure the robustness of the proposed algorithm, the normalized correlation (NC) is calculated between the original watermark image and the extracted watermark image. The NC value is calculated using Equation (20):

$$\mathrm{NC}(W, W^*) = \frac{\sum_{k=1}^{N}\sum_{l=1}^{N} w(k,l)\cdot w^*(k,l)}{\sqrt{\sum_{k=1}^{N}\sum_{l=1}^{N} w(k,l)\cdot w(k,l)}\sqrt{\sum_{k=1}^{N}\sum_{l=1}^{N} w^*(k,l)\cdot w^*(k,l)}} \tag{20}$$

where W and W^* are the original watermark and extracted watermark, respectively.

Now, the main fact is to consider the different types of noise attacks on the watermarked image. The results are illustrated in such a way as to identify the effect of keys on the NC values. Figures 7–9 show the effect of keys in a pictorial way, including the PSNR and NC values.

Attack type		Lena	Peppers	Baboon	Fruit
No attack	Watermarked image				
	PSNR	50.04	49.78	51.56	52.64
	Extracted watermark (with and without keys)				
	NC (with and without keys)	1.0 and 1.0	1.0 and 1.0	1.0 and 1.0	1.0 and 1.0
Gaussian noise (0.1)	Watermarked image				
	Extracted watermark (with and without keys)				
	NC (with and without keys)	1.0 and 0.9997	1.0 and 0.9823	1.0 and 1.0	1.0 and 0.9351
Speckle noise (0.01)	Watermarked image			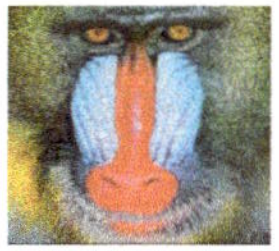	
	Extracted watermark (with and without keys)				
	NC (with and without keys)	1.0 and 0.8835	1.0 and 0.9292	1.0 and 0.9068	1.0 and 0.9349

Figure 7. *Cont.*

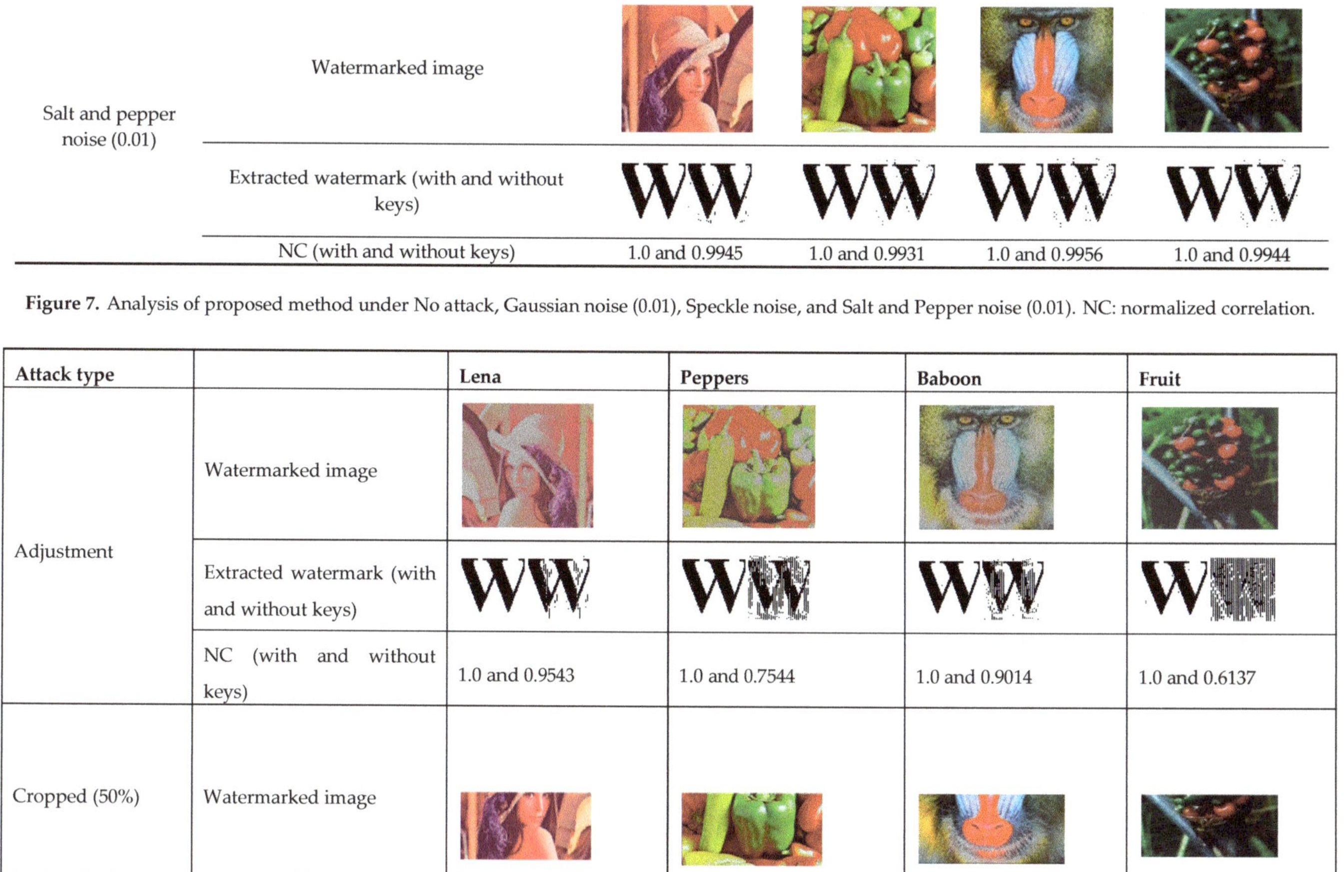

Salt and pepper noise (0.01)	Watermarked image				
	Extracted watermark (with and without keys)				
	NC (with and without keys)	1.0 and 0.9945	1.0 and 0.9931	1.0 and 0.9956	1.0 and 0.9944

Figure 7. Analysis of proposed method under No attack, Gaussian noise (0.01), Speckle noise, and Salt and Pepper noise (0.01). NC: normalized correlation.

Attack type		Lena	Peppers	Baboon	Fruit
Adjustment	Watermarked image				
	Extracted watermark (with and without keys)				
	NC (with and without keys)	1.0 and 0.9543	1.0 and 0.7544	1.0 and 0.9014	1.0 and 0.6137
Cropped (50%)	Watermarked image				

Figure 8. *Cont.*

	Extracted watermark (with and without keys)				
	NC (with and without keys)	1.0 and 0.7919	1.0 and 0.7821	1.0 and 0.7912	1.0 and 0.7866
Sharpening (tolerance = 0.1)	Watermarked image				
	Extracted watermark (with and without keys)				
	NC (with and without keys)	1.0 and 0.9578	1.0 and 0.9335	1.0 and 0.9241	1.0 and 0.8594
Weiner filtering	Watermarked image				
	Extracted watermark (with and without keys)				
	NC (with and without keys)	1.0 and 0.6753	1.0 and 0.6785	1.0 and 0.6884	1.0 and 0.6771

Figure 8. Analysis of proposed method under Adjustment, Cropping (50%), Sharpening (0.1), and Weiner filtering.

Attack type		Lena	Peppers	Baboon	Fruit
Poison noise	Watermarked image				
	Extracted watermark (with and without keys)				
	NC (with and without keys)	1.0 and 0.9950	1.0 and 0.9963	1.0 and 0.9992	1.0 and 0.9990
Median filtering	Watermarked image				
	Extracted watermark (with and without keys)				
	NC (with and without keys)	1.0 and 0.9762	1.0 and 0.9541	1.0 and 0.9896	1.0 and 0.9459
Compression (quality factor: 50%)	Watermarked image				
	Extracted watermark (with and without keys)				
	NC (with and without keys)	1.0 and 0.5775	1.0 and 0.5936	1.0 and 0.5912	1.0 and 0.5676

Figure 9. *Cont.*

Rotation (40^0)	Watermarked image			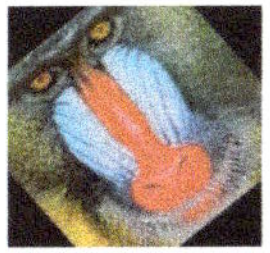	
	Extracted watermark (with and without keys)				
	NC (with and without keys)	1.0 and 0.5160	1.0 and 0.5132	1.0 and 0.5194	1.0 and 0.5193

Figure 9. Analysis of the proposed method under Poison noise, Median filtering, Compression (quality factor: 50%), and Rotation.

Furthermore, Tables 2 and 3 show an overview of the NC values of the proposed scheme with keys and without keys, respectively. Notably, the NC values shown in Table 2 reflect better results than Table 3. This is because severe noise attacks affect the degree of ascendant/descendant (*dof*). This *dof* is derived without key $k1$ and is vulnerable to noise attack until it is modified with dof', as defined in Equations (14) and (15). Since the keys make the system effectively resistant against noise, Table 3 shows better results in terms of NC. Further, the extracted watermarks from four different watermarked images under Gaussian noise with tolerance 0.1 are shown in Figure 7. The extracted watermark using only *dof* (without key) for the "Fruit" cover image provides lower NC values than the others. Since the color variation in this host image is not very high, *dof* (without keys) is more vulnerable under additive noise attack. This is also applicable for other attacks, such as adjustment and sharpening, as shown in Figure 8. This problem is overcome in the proposed framework with the concept of key mapping. In spite of the high noise attack, extracting the watermark using *dofk* (with keys) could reconstruct the watermark image successfully with a unity of NC values, as shown in Figures 7–9. We observed that the NC of the proposed method against various attacks is numerically one. It is because the keys ($k1$, $k2$ and $k3$) that contain the necessary information of the watermark are not affected by various attacks. Hence, this proposed technique ensures high robustness.

Table 2. NC values after applying various noise attacks (with keys).

No	Attack Type	Lena	Peppers	Baboon	Fruit
1	Gaussian (0.01)	1.0	1.0	1.0	1.0
2	Speckle (0.01)	1.0	1.0	1.0	1.0
3	Adjustment	1.0	1.0	1.0	1.0
4	Cropping (50%)	1.0	1.0	1.0	1.0
5	Sharpening (tol = 0.1)	1.0	1.0	1.0	1.0
6	Rotation (40^0)	1.0	1.0	1.0	1.0
7	Wiener filtering	1.0	1.0	1.0	1.0
8	Poison noise	1.0	1.0	1.0	1.0
9	Salt and pepper noise (0.01)	1.0	1.0	1.0	1.0
10	Median filtering	1.0	1.0	1.0	1.0
11	Compression (quality factor = 50%)	1.0	1.0	1.0	1.0

Table 3. NC values after applying various noise attacks (without keys).

No	Attack Type	Lena	Peppers	Baboon	Fruit
1	Gaussian (0.1)	0.9997	0.9823	1.0	0.9351
2	Speckle (0.01)	0.8835	0.9292	0.9068	0.9349
3	Adjustment	0.9543	0.7544	0.9014	0.6137
4	Cropping (50%)	0.7919	0.7821	0.7912	0.7866
5	Sharpening (tol = 0.1)	0.9578	0.9335	0.9241	0.8594
6	Rotation (40^0)	0.5160	0.5132	0.5194	0.5193
7	Wiener filtering	0.6753	0.6785	0.6884	0.6771
8	Poison noise	0.9950	0.9963	0.9992	0.9990
9	Salt and pepper noise (0.01)	0.9945	0.9931	0.9956	0.9944
10	Median filtering	0.9762	0.9541	0.9896	0.9459
11	Compression (quality factor = 50%)	0.5775	0.5936	0.5912	0.5676

Table 4 shows a comparative analysis between the proposed and several recent state-of-the-art methods [13,23,24] for NC against different attacks. From this table, it is observed that the NC of the proposed method is numerically one against various attacks using keys, in contrast to state-of-the-art methods whose NC vary from 0.7991 to 0.9999. It should be mentioned that Ahmed et al. [23] shows low robustness against rotation and salt and pepper noise attack, and Su et al. [13] shows low robustness against median filtering and JPEG compression attack. In all other cases, these two methods show good robustness. Moreover, Patvardhar et al. [24] shows good robustness against various attacks.

Table 4. A comparative analysis between the proposed and several recent methods in terms of NC.

No	Attack Type	Ahmed et al. [23]	Patvardhar et al. [24]	Su et al. [13]	Proposed
1	Gaussian noise (0.1)	0.9625	0.9885	0.9131	1.0
2	Speckle noise (0.01)	0.9601	–	–	1.0
3	Contrast Adjustment	–	0.9491	–	1.0
4	Cropping (50%)	–	0.9947	0.9604	1.0
5	Sharpening	0.9388	–	0.9999	1.0
6	Rotation (25°)	0.7991	0.9989	–	1.0
7	Poison noise	0.9884	–	–	1.0
8	Salt and pepper noise (0.01)	0.9117	0.9807	0.9902	1.0
9	Median filtering	0.9908	0.9989	0.8814	1.0
10	JPEG compression (quality factor = 20%)	0.9784	0.9895	0.8469	1.0

In other words, this proposed algorithm with its unique key approach is much more robust than any other existing method. In addition, our method utilizes the key mapping concept with singular values of the host image. This concept improves the performance of the proposed method against severe noise attack. This approach also ensures ownership with high robustness. Furthermore, Gaussian mapping enhances the security of the watermark. Finally, coefficient ordering in the smaller block provides high imperceptibility. The concatenation of smaller blocks into the larger block provides high robustness against noise attack. In a nutshell, it can be concluded that our proposed method outperforms recent state-of-the-art methods in terms of robustness, security, and imperceptibility.

5. Conclusions

This paper presented an image watermarking scheme using FWHT, SVD, key mapping, and coefficient ordering. FWHT is chosen because of its low computational complexity. To enhance the robustness of the proposed method against severe attacks, key mapping is introduced using SVD. It is used because unique keys are generated from the singular values of the FWHT blocks of the cover image. Furthermore, Gaussian mapping is used to scramble the watermark. This makes the system secured against unauthorized detection. Thus, the proposed method ensures high robustness as well as high security against numerous attacks. Experimental results indicated that the proposed scheme shows better results than the recent methods in terms of robustness and security. Moreover, it yielded high-quality watermarked images. The NC value of the proposed method is numerically one, while the PSNR of it lies between 49.78 and 52.64. In contrast, the recent state-of-the-art methods show that the NC varies from 0.7991 to 0.9999, while the PSNR resides between 39.4428 and 54.2599. These results verified that the proposed method could be effectively utilized for image copyright protection and proof of ownership. We will extend the proposed method for video watermarking in the future.

Author Contributions: All authors contributed equally to the conception of the idea, the design of experiments, the analysis and interpretation of results, and the writing and improvement of the manuscript. All authors have read and agreed to the published version of the manuscript.

Funding: This work was supported by the Korea Institute of Energy Technology Evaluation and Planning (KETEP) and the Ministry of Trade, Industry and Energy (MOTIE) of the Republic of Korea (20192510102510, 20172510102130)

Conflicts of Interest: The authors declare no conflict of interest.

References

1. Bakhsh, F.Y.; Moghaddam, M.E. A robust HDR images watermarking method using artificial bee colony algorithm. *J. Inf. Secur. Appl.* **2018**, *41*, 12–27.
2. Guo, Y.; Au, O.C.; Wang, R.; Fang, L.; Cao, X. Halftone image watermarking by content aware double-sided embedding error diffusion. *IEEE Trans. Image Process.* **2018**, *27*, 3387–3402. [CrossRef] [PubMed]
3. Chetan, K.R.; Nirmala, S. An efficient and secure robust watermarking scheme for document images using Integer wavelets and block coding of binary watermarks. *J. Inf. Secur. Appl.* **2015**, *24*, 13–24. [CrossRef]

4. Wanga, H.; Yina, B.; Zhoub, L. Geometrically invariant image watermarking using connected objects and gravity centers. *KSII Trans. Internet Inf. Syst.* **2013**, *7*, 2893–2912.
5. Lin, P.Y.; Chen, Y.H.; Chang, C.C.; Lee, J.S. Contrast adaptive removable visible watermarking (CARVW) mechanism. *Image Vis. Comput.* **2013**, *31*, 311–321. [CrossRef]
6. Su, Q.; Niu, Y.; Zou, H.; Liu, X. A blind dual color images watermarking based on singular value decomposition. *Appl. Math. Comput.* **2013**, *219*, 8455–8466. [CrossRef]
7. Wu, X.; Sun, W. Robust copyright protection scheme for digital images using overlapping DCT and SVD. *Appl. Soft Comput.* **2013**, *13*, 1170–1182. [CrossRef]
8. Bhatnagar, G.; Wua, Q.M.J.; Raman, B. A new robust adjustable logo watermarking scheme. *Comput. Secur.* **2012**, *31*, 40–58. [CrossRef]
9. Tsai, H.H.; Huang, Y.J.; Lai, Y.S. An SVD-based image watermarking in wavelet domain using SVR and PSO. *Appl. Soft Comput.* **2012**, *12*, 2442–2453. [CrossRef]
10. Hun, H.T.; Chen, W.H. A dual cepstrum based watermarking scheme with self-synchronization. *Signal Process.* **2012**, *92*, 1109–1116.
11. Lin, C.C. An information hiding scheme with minimal image distortion. *Comput. Stand. Interfaces* **2011**, *33*, 477–484. [CrossRef]
12. Lee, Y.; Kim, H.; Park, Y. A new data hiding scheme for binary image authentication with small image distortion. *Inf. Sci.* **2009**, *179*, 3866–3884. [CrossRef]
13. Su, Q.; Wang, G.; Zhang, X. A new algorithm of blind color image watermarking based on LU decomposition. *Multidimens. Syst. Signal Process.* **2018**, *29*, 1055–1074. [CrossRef]
14. Murty, P.S.; Kumar, S.D.; Kumar, P.R. A semi blind self reference image watermarking in DCT using Singular Value Decomposition. *Int. J. Comput. Appl.* **2013**, *62*, 29–36.
15. Shen, J.J.; Ren, J.M. A robust associative watermarking technique based on vector quantization. *Digit. Signal Process.* **2010**, *20*, 1408–1423. [CrossRef]
16. Bhatnagar, G.; Raman, B. A new robust reference watermarking scheme based on DWT-SVD. *Comput. Stand. Interfaces* **2009**, *31*, 1002–1013. [CrossRef]
17. Zhou, X.; Zhang, H.; Wang, C. A robust image watermarking technique based on DWT, APDCBT, and SVD. *Symmetry* **2018**, *10*, 77. [CrossRef]
18. Sarker, M.I.H.; Khan, M.I. An efficient image watermarking scheme using BFS technique based on Hadamar Transform. *Smart Comput. Rev.* **2013**, *3*, 298–308.
19. Kumar, A.; Luhach, A.K.; Pal, D. Robust digital image watermarking technique using image normalization and Discrete Cosine Transformation. *Int. J. Comput. Appl.* **2013**, *65*, 5–13.
20. Liua, J.; Liub, G.; Hea, W.; Lia, Y. A new digital watermarking algorithm based on WBCT. *Procedia Eng.* **2012**, *29*, 1559–1564. [CrossRef]
21. Lai, C.C.; Tsai, C.C. Digital image watermarking using Discrete Wavelet Transform and Singular Value Decomposition. *IEEE Trans. Instrum. Meas.* **2010**, *59*, 3060–3063. [CrossRef]
22. Mohammad, A.A.; Alhaj, A.; Shaltaf, S. An improved SVD-based watermarking scheme for protecting rightful ownership. *Signal Process.* **2008**, *88*, 2158–2180. [CrossRef]
23. Ahmed, K.A.; Ozturk, S. A novel hybrid DCT and DWT based robust watermarking algorithm for color images. *Multimed. Tools Appl.* **2019**, *78*, 17027–17049.
24. Patvardhan, C.; Kumar, P.; Lakshmi, C.V. Effective color image watermarking scheme using YCbCr color space and QR code. *Multimed. Tools Appl.* **2018**, *77*, 12655–12677. [CrossRef]

Article

Hierarchical Intrusion Detection Using Machine Learning and Knowledge Model

Martin Sarnovsky * and Jan Paralic

Department of Cybernetics and Artificial Intelligence, Faculty of Electrical Engineering and Informatics, Technical University of Košice, Letna 9, 040 01 Košice, Slovakia; jan.paralic@tuke.sk
* Correspondence: martin.sarnovsky@tuke.sk

Received: 31 December 2019; Accepted: 25 January 2020; Published: 1 February 2020

Abstract: Intrusion detection systems (IDS) present a critical component of network infrastructures. Machine learning models are widely used in the IDS to learn the patterns in the network data and to detect the possible attacks in the network traffic. Ensemble models combining a variety of different machine learning models proved to be efficient in this domain. On the other hand, knowledge models have been explicitly designed for the description of the attacks and used in ontology-based IDS. In this paper, we propose a hierarchical IDS based on the original symmetrical combination of machine learning approach with knowledge-based approach to support detection of existing types and severity of new types of network attacks. Multi-stage hierarchical prediction consists of the predictive models able to distinguish the normal connections from the attacks and then to predict the attack classes and concrete attack types. The knowledge model enables to navigate through the attack taxonomy and to select the appropriate model to perform a prediction on the selected level. Designed IDS was evaluated on a widely used KDD 99 dataset and compared to similar approaches.

Keywords: intrusion detection; machine learning; classification; knowledge modelling

---

1. Introduction

With the increasing extent, diversity and added value of information and communication systems (ICT) usage, especially in today's networked world, the number and types of security attacks increase also (see, e.g., [1]). Companies, as well as public institutions, invest constantly growing parts of their ICT budgets on their network and computer security. In order to efficiently cope with this situation, research in the ICT security-related domain is very much needed.

We can observe two main approaches to IDS in the research literature. One focuses more on machine learning approaches with the aim to achieve the best possible prediction of attacks, or even specific attack types. The other one (knowledge-based approach) is more user-centric and tries to model the necessary background knowledge in the network security domain. This article presents a particular and specific contribution to the mentioned challenges in terms of knowledge-based IDS design, implementation and experimental evaluation. The main motivation is to get the best out of both worlds, i.e., extremely good performance in terms of prediction accuracy and false alarm rates, on one hand, and proper coverage of the relevant context as well as the provision of proper additional information to the network operators, on the other. We build on the current state-of-the-art in the two prominent areas in IDS, contributing to ICT security-threat prevention, i.e., machine learning and knowledge-based approaches. We are looking for symmetry in using both of these principal approaches, which means that both of them are equally important and mutually contribute to addressing intrusion detection and prevention challenges.

The rest of this paper is organized as follows. Related work is presented in the following Section 2. Main building blocks of this symmetry-following strategy are outlined in Section 3.1, followed by the detailed description of the two main building blocks, namely the knowledge model (described in Section 3.2) and machine learning approaches (Section 3.3). The mutual cooperation of the two main building blocks of our solution is presented in Section 3.4.

Section 4 presents the experimental evaluation of the proposed symmetry-following system for intrusion detection. We start with the introduction of performance metrics used (Section 4.1) in the subsequent experiments. Experimental evaluation is divided into three parts. We first present the performance of the two main prediction models of our hierarchical IDS in Section 4.2.1, continue with the explanation of the interplay between our knowledge model and prediction models in Section 4.2.2 and performance of a specific prediction model targeted for seldom new attack types in Section 4.2.3. Finally, Section 5 summarizes the main findings from our experiments and highlights the advantages of the proposed solution, which brings the symmetry-following strategy.

2. Related Work

In recent years, the use of machine learning methods in network intrusion detection domain became a widely studied area of research. With several publicly available datasets, numerous different approaches emerged to tackle the problem of detection of network attacks. Currently, popular methods used in the intrusion detection domain are neural networks models. As they usually gain better performance than traditional machine learning models, they became popular methods in this area, although their explanatory capacity is often very limited. Numerous different approaches are used, including both shallow and deep learning methods [2], or a combination of both [3].

Standalone data analytics and machine learning methods are often combined into more advanced hybrid approaches, combining various methods including the use of hierarchical and layered models [4–7] and anomaly detection models [8,9] or enhancing the machine learning models with knowledge-based approaches [10,11]. The overall motivation of those approaches is usually aimed to provide the capability of detection of rare attacks without sacrificing the detection accuracy of the frequent ones and, on the other hand, minimizing the false alarm rate in such attacks. Ahmin et al. [12] proposed a hierarchical model combining tree and rule-based concepts and used three models (REP Tree, JRip, Forest PA), where the outputs of the two models are used as the inputs to the third one, creating a hierarchical model consisting of three classifiers, predicting either normal or attack connections and attack types. The approach was evaluated on the CICIDS 2017 dataset and achieved an accuracy of 96.65%, while keeping detection rate at 94,47%. Compared to other models on the same dataset, this approach gained superior accuracy performance, while still keeping reasonable training and evaluation time. In Reference [13], a layered-approach with multi-classifiers is presented. To improve the classification of minority attacks, authors used a combination of Naive Bayes and NBTree models on the NSL-KDD dataset. The IDS uses four layers, the top two for detection of major attacks (e.g., DoS, Probe) and the other two layers for minority attack classes (U2R, R2L). The model uses different reduced feature sets for each attack prediction; authors kept only a subset of relevant features for each attack class using domain knowledge and continuous experiments with the models. Evaluation proved an improvement in classification in minor classes compared to standard approaches, gaining overall accuracy of 99.05%. In Reference [14], Gain Ratio was used to select the features for each layer of the layered model consisting of trees, Naive Bayes and Perceptron models. In Reference [15], a similar goal was achieved using Conditional Random Fields approach. Recently, Zhou et al. [16] presented the ensemble model consisting of C4.5, Random Forest and PA Forest, where the decisions are based on the average of probabilities rule (voting based on average of probabilities). The ensemble model uses feature selection by applying a hybrid approach of combination of correlation-based feature selection and biologically inspired bat algorithm. The model was tested on KDD 99, CICIDS 2017 and NSL-KDD datasets, achieving superior performance compared to other similar models (accuracy of 99,9% on KDD, 96,8% on CICIDS 2017 and 99,1% on NSL-KDD).

In the area of knowledge models for intrusion detection systems, there have been ontologies explicitly designed for the description of the network attacks [10,17,18], more specifically targeted to specific issues related to the domain, e.g., vulnerabilities [19] or, on the other hand, covering a broader range, e.g., entire cybersecurity [20]. Such ontologies were in the past used to develop the ontology-based IDS. In Reference [21], authors present the DAML+OIL attack ontology and corresponding detection system applied on the KDD 99 data. The ontology includes high-level domain concepts (e.g., Attack, Host, Component, etc.) and defines the basic relations between the protocols (Components) and attacks, organized in the taxonomy. Abdoli et al. [22] use the ontology to capture the semantic relations between the attacks in distributed IDS and improve the detection using such knowledge. Authors studied the attack scenarios and their effects and identified the attack relationships, which were incorporated into the ontology. Ontology was evaluated on KDD 99 dataset, focusing on the DoS attack classes, recognizing 99.9% of DoS attacks in the testing data. In distributed IDS, ontologies were also used to support cooperative detection [23]. In Reference [24], ontology-based, situation-aware intrusion detection system is presented. The approach is based on the integration of various heterogeneous data sources used to create the semantically rich knowledge base. Reasoner processes the streams from the sources, extracts the knowledge and uses the ontology to infer the possibility of an attack. More recently, OWL-S attack ontology was used in IDS to detect the protocol-specific attacks [25], which was evaluated on the KDD 99 dataset. In Reference [26], authors describe the usage of the ontology in semantic intrusion detection system for malware detection.

The approach presented in this paper combines both studied areas—hybrid hierarchical classifiers with knowledge-based approaches. Hierarchical classification is based on the taxonomy of the attack classes and uses various classification models to handle the prediction on a specific level of the target attribute. To predict the particular class of attack, we used an ensemble model with weighted voting for particular classes to achieve more robust classification even in minority classes. The entire process is guided by the ontology, which is used to navigate through the attack taxonomy and retrieve the corresponding model to perform the prediction and domain-specific knowledge related to the severity of the attacks, which could complement the predictions.

3. Hierarchical Intrusion Detection

3.1. The Overall Architecture of the Proposed System

The main objective of the proposed approach is to combine the multi-layered hierarchical model approach of using various predictive models to detect the intrusions on different levels of the attack type taxonomy and combine it with the knowledge obtained from the domain-specific knowledge contained in the ontology. Scheme depicted in Figure 1 outlines how the knowledge model is used in the prediction.

Hierarchical classification can be divided into two separate phases.

1. Normal/Attack separation—the first phase is a binary classification task. The classifier used in this phase is used to distinguish normal traffic and attacks. If a connection is labelled as a normal one, then an alarm is not raised. Otherwise, the suspicious connection is processed by a set of models to determine the class of attack during the phase 2.
2. Attack class and type prediction—this phase is guided by the taxonomy of the attacks from the knowledge model. The system hierarchically processes the taxonomy and selects the appropriate model to classify the instance on a particular level of a class hierarchy.
3. When a class of attack is predicted, ontology is queried for all relevant sub-types of the attack type and to retrieve the suitable model to predict the particular sub-type. Knowledge model can also be used to extract specific domain-related information as a new attribute, which could be used either to improve the classifier's performance or to provide context, domain-specific information which could complement the predictive model.

More in-detail description of the predictive models and their evaluation will be provided in the following sub-sections.

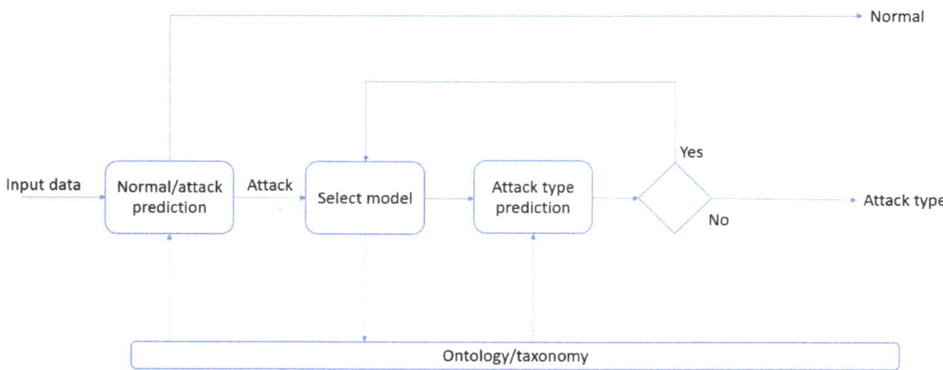

Figure 1. The overall architecture of the proposed intrusion detection system.

3.2. Network Intrusion Knowledge Model

The central part of the knowledge model is the attack taxonomy. Figure 2 illustrates the overall taxonomy of the network attack included KDD 99 dataset, which is one of the (still) most frequently used data collections in the selected domain [27,28]. Attacks are divided into four groups (DOS, R2L, U2R, Probe), each of them including specific attack types.

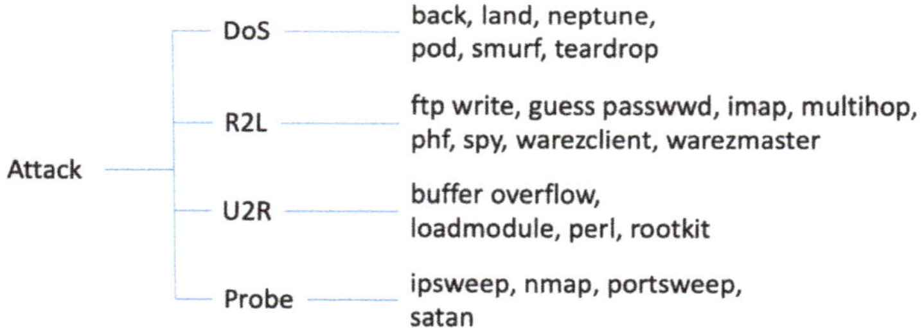

Figure 2. Taxonomy of the target attribute in KDD 99 dataset.

Besides the taxonomy, the knowledge model also covers the domain-specific knowledge from the network intrusion field. Basic concepts and relations were extracted from the already existing security domain models and standards [29]. As the objective of the knowledge model was to use it in the data analytical tasks, the concepts and properties had to be able to be mapped to the data used in the process. Moreover, concepts related to the classification models had to be included, to create the relation between the particular classifier and its usability on the specific level of target attribute hierarchy. Figure 3 visualizes the top-level concepts and relations between them.

- Connection class represents particular connections, whether normal ones or attacks. The class forms a class hierarchy when a sub-class Attacks represents the attack. Attack sub-classes (TypeOfAttack) represent the classes of the attacks (e.g., DoS, r2l, etc.); concrete attacks types are modelled as a sub-class of the ConcreteAttack type classes (e.g., back, land, etc.).
- Effect class covers all possible effects that an attack affects (e.g., slowing down of the server response, gaining root access for the user, service outage, etc.)
- Mechanism class and its sub-classes describe all possible mechanisms of particular attacks (e.g., poor environment maintenance, incorrect configuration of the components, etc.)

- Flag characterizes the normal or error states of the specific connections (e.g., service not responding, denied the connection, etc.)
- Protocol represents the protocols used in the connection (e.g., TCP, UDP, etc.)
- Service concept describes service types related to the connection (e.g., http, telnet, etc.)
- Severity describes how severe the possible attack type effects could be (low, medium and high).
- Targets define the possible targets of the particular attack type (e.g., user, network, data, etc.).
- Models concept covers the classification models used to predict the given target attribute

Ontology was implemented in OWL (Web Ontology Language). Particular ontology instances represent the particular connections (e.g., connection records from the data). Trained and serialized classification models are instantiated as the instances of the *Model* class. The models could be accessed using their *URI* property, which contains the URI of the stored serialized model in the system.

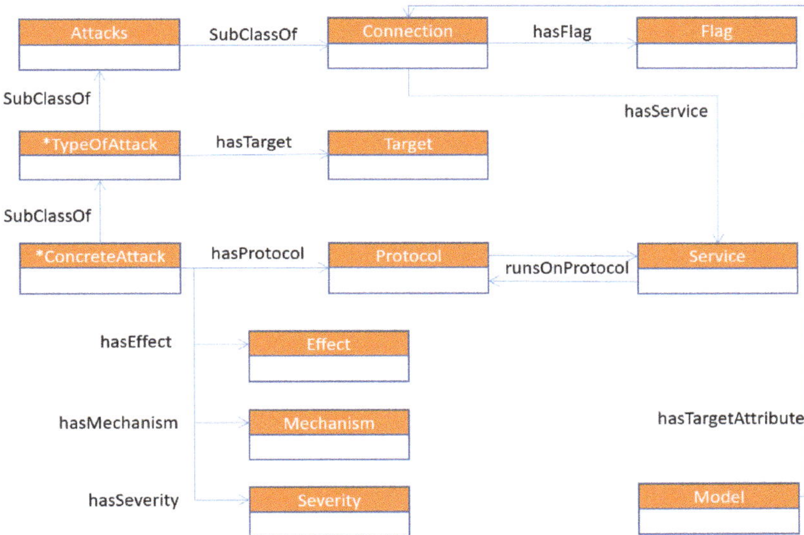

Figure 3. Knowledge model for the network intrusion detection domain.

3.3. Machine learning Models for Detection of the Network Attacks on KDD 99 Dataset

To evaluate the proposed approach, we used the KDD Cup 1999 competition dataset 1999 [30], a commonly used network intrusion detection data collection. Dataset consists of the device logs in a LAN network collected over nine weeks. We used the 10% sample subset of the dataset, which consisted of 494,021 records. Each record represents a connection and is labelled as a normal one, or as an attack (exactly one attack type assigned). The target attribute (attack type) can be organized in a concept hierarchy, which is specified in the knowledge model (see Figure 2). There are 24 different attack types present in the dataset, and each attack type belongs to one of the four attack classes. The target attribute is heavily imbalanced, Table 1. summarizes the number of records of attack types and attack classes.

Table 1. Attack types and classes.

Attack	Attack Class	Number of Samples
back	DoS	2203
land		21
neptune		107,201
pod		264
smurf		280,790
teardrop		979
satan	Probe	1589
ipsweep		1247
nmap		231
portsweep		1040
guess_passwd	R2L	53
ftp_write		8
imap		12
phf		4
multihop		7
warezmaster		20
warezclient		1020
spy		2
buffer_overflow	U2R	30
loadmodule		9
perl		3
rootkit		10
normal	Normal	97,227

Each connection is described using a set of features: basic features of the connection, content features and traffic features (overall 32 features). The first group describes the protocol type, duration of the connection, service on the destination of the attack and other features relevant to the standard TCP connection description. Content features are attributes that could be linked to the domain-specific knowledge. Traffic features describe the two-second time window attributes, e.g., a number of connections to the same host in such window. During the pre-processing, we mostly focused on feature selection. According to the work of Zhou and Cheng in [16], we selected only the most relevant attributes: service, src_bytes, dst_bytes, logged_in, num_file_creations, srv_count, serror_rate, rerror_rate, srv_diff_host_rate, dst_host_count, dst_host_diff_srv_rate, dst_host_srv_diff_host_rate.

To illustrate the proposed approach on the KDD 99 network intrusion dataset, Figure 4 depicts the structure of the prediction models used on different levels of the target attribute class hierarchy. During the first phase, an *attack detection model* is used for prediction on the top-level of the class hierarchy, e.g., to distinguish between the attack connections from the normal ones. The classifier was trained on all instances of the dataset and target attribute was transformed to a binary one reflecting the class hierarchy. The main goal of the classifier is to reliably predict the normal connections and separate them from the attack ones.

If the model detects an attack connection, the *ensemble model* is used to predict one of the four types of the attack. In this case, we used an ensemble of classifiers with a weighted voting scheme, trained on all attack instances in the dataset. The main reason to use such an approach was the effect of class imbalance on this level. Standard machine learning models were able to gain good accuracy, achieved mostly by good performance on the dominant class (in this case DoS attacks). However, the models struggled to predict minor classes such as U2R.

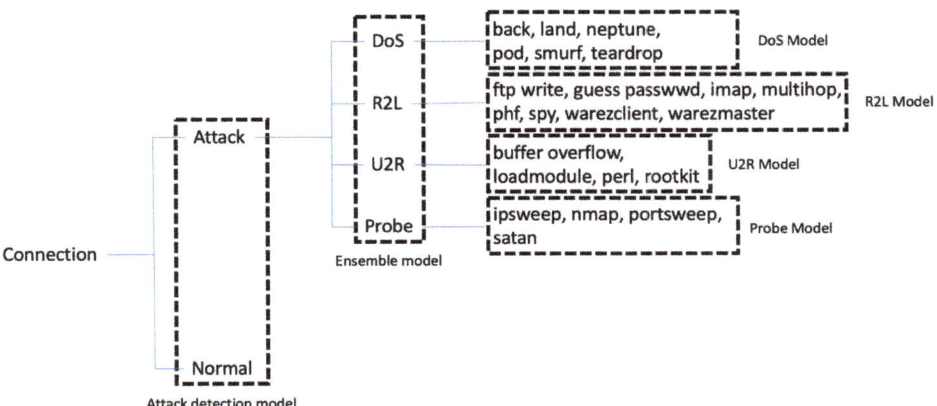

Figure 4. Different models used for predictions on different levels of target attribute class hierarchy.

Some standard classifiers performed very well when predicting the specific classes but failed (or produced significant errors) on the other ones. For example, when training a decision tree model on the attack connections, the model performed very well when classifying the DoS and R2L classes, on the other hand, missed an almost significant amount of the Probing attacks and was not able to detect the U2R attacks at all. That leads us to the idea of complementing such classifier with a set of other models, able to reliably predict the other classes. To perform a final prediction, weighted voting could be implemented, based on the classification performance in prediction of a particular class. The weighting scheme is depicted in Table 2.

Table 2. Weighting scheme used in the ensemble model.

Weighting Scheme	Class 1	Class 2	Class 3	Class 4
model 1	$w_{1,1}$	$w_{1,2}$	$w_{1,3}$	$w_{1,4}$
model 2	$w_{2,1}$	$w_{2,2}$	$w_{2,3}$	$w_{2,4}$
model 3	$w_{3,1}$	$w_{3,2}$	$w_{3,3}$	$w_{3,4}$
...	...	...	...	...

After the prediction of the attack class, particular models were employed to predict the concrete type of the specific attack class. Four different models were trained, each for prediction of the particular attack types for each attack class. The models were trained using only instances of particular attack class records from the dataset. This proved to be difficult on minority class (U2R), as the dataset contains very few records of that type.

Then, when an attack class is predicted, a set of models are dedicated to predicting the concrete attack type (e.g., DoS model to predict the concrete type of DoS attack, when such attack is detected by the IDS). There are four different models trained, each dedicated to prediction of each attack class' types.

All models were trained in the Python environment using standard packages such as scikit-learn in Python. Predictive models are then serialized using native or supporting tools (e.g., Pickle in Python). Serialized models URI (Uniform Resource Identifier) was then passed to the knowledge model.

3.4. Use of Knowledge Model in Multi-Stage Intrusion Detection

Implemented semantic model is used in the hierarchical classification mostly to navigate through the target class taxonomy and to select the appropriate model to perform a prediction on the chosen level. The detection system is implemented in Python and communicates with the ontology using RDFlib package, which enables to specify the SPARQL queries and retrieve the results as a Python object. When predicting the unknown connection, the system checks the ontology taxonomy and starts from the top level. Using SPARQL query, it checks the top level of the class hierarchy and retrieves the corresponding model for given prediction type using the hasTargetAttribute property (e.g., attack detection). As the classification models are serialized and stored, they can be identified and retrieved using their URIs. Once a classifier is retrieved, it is deserialized in the system and used to predict the given unknown example. Once the prediction is performed, the system checks the prediction and queries the ontology to see if there is a classifier able to process the record further (e.g., to predict the type of the attack) and retrieves the corresponding classifier.

As the prediction of concrete sub-type of the attacks could be in some cases tricky (minority classes), we considered a retrieval of the information from the ontology, which could be beneficial without predicting a correct attack type. Moreover, in the network intrusion domain, new attacks (attacks, that were not present in the training data) can appear when using the models. In such a case, the system is able to predict the attack class and use domain-specific information from the knowledge model to enhance the prediction with domain-specific knowledge, e.g., severity of the detected attack. Besides using the class hierarchy, we experimented by using the domain-specific knowledge from the ontology to improve the detection system. We decided we can leverage the expert information about the possible effects of the given attacks and their severity. If the model is not reliable enough to predict the concrete attack sub-type, the system can be used to provide at least information on how severe the attack would be. Using the Effect class and hasSeverity property, we were able to retrieve the severity values of the particular attacks and enhance the dataset with this newly acquired attribute. We experimented then with the training of the classifiers able to predict the severity of the attacks, which could serve as a supporting source of information, complementing the attack type prediction or providing the severity information.

4. Experimental Evaluation

4.1. Performance Metrics

To evaluate the performance of the designed approach, we used metrics commonly employed in classification tasks such as precision and recall. To evaluate the prediction of the particular classes, we used the confusion matrix. The matrix is essential, as it precisely depicts both, correctly and incorrectly classified records. Especially in case of not balanced target classes, we could estimate the classification errors. When predicting the network attacks on the first level of target taxonomy, we used the confusion matrix for binary classification and measured the standard evaluation metrics:

Precision:

$$P = \frac{TP}{TP + FP}, \tag{1}$$

Recall:

$$R = \frac{TP}{TP + FN}, \tag{2}$$

where:

- TP (True Positive): when predicted network attack is in fact an attack,
- TN (True Negative): when predicted normal record is in fact normal record,
- FN (False Negative): when predicted normal record is in fact an attack,
- FP (False Positive): when predicted network attack is in fact a normal record.

From the perspective of the performance evaluation of the entire network intrusion detection, we decided to evaluate the models also from the perspective of missed attacks and raised false alarms. To evaluate such aspects, F-measure combining both, precision and recall metrics can be used, as well as FAR (False Alarm Rate), which is the total number of incorrectly identified attacks (FP) divided by the total number of normal records (TN + FP).

Precision and recall metrics could be directly used to evaluate the first level of hierarchical classification, as the target attribute was a binary one. In the following stages, we computed the precision and recall metrics for each target attribute value and to measure the overall precision and recall; we used macro-averaging. This was more important, as it was crucial to precisely see the number of false alarms and missed attacks to correctly estimate the model quality. Then, we used a multi-class confusion matrix to estimate the overall classifier performance.

4.2. Performance Evaluation

4.2.1. Model Training and Evaluation

For the attack detection model, we used a classifier based on decision trees. Dataset consisted of all dataset records; target attribute was transformed to a binary one, labelling the normal and attack connections. We split the dataset in a 70/30 training/testing ratio. The classifier was trained without limitation of maximum depth of the tree and used gini as a splitting criterion. The testing data were not used in the training of any of the models (ensemble and attack type models) and were also used to evaluate the whole system. Attack detection model was evaluated on the testing set, and the model gained accuracy of 0.9997; confusion matrix is depicted in Table 3.

Table 3. Results of the decision tree classifier for attack prediction (attack detection model).

Attack Detection Model	Normal	Attack	Precision	Recall
Normal	29,095	11	0.999	0.999
Attack	35	119,066		

An *ensemble classification model* was trained using the attack records from the dataset (97277 normal records were removed from training). In the ensemble model, we used Naive Bayes and Decision Tree classifiers as the base models. We chose the different tree and Naive Bayes models in order to have models that correctly cover all attack classes in the dataset. When training tree-based models, experiments proved, that such models were performing reasonably on Probe, DoS and R2L attacks. On the other hand, U2R (minor class) class produced either many false alarms, or the models were not able to detect the U2R attacks at all. Therefore, separate one-vs-all models were trained to detect just the U2R attacks. Ensemble model was then able to leverage both classifier types—tree-based model to detect the R2L, Probe and DoS attacks and one-vs-all model to detect the U2R attacks. The training set consisted only of attack connections (396743 records) and the models were trained and evaluated on such data using the 70/30 training/testing split and evaluated their performance on the particular classes. Class accuracies were then used to set the weights of the ensemble model created of these base classifiers. Then, we evaluated the overall ensemble model on the testing set of the KDD 99 dataset (we did not use the records used in training). Table 4 depicts the confusion matrix of the ensemble model performance when classifying the attack records.

Table 4. Confusion matrix of the ensemble model for attack type prediction (ensemble classification model).

Ensemble Model	Probe	U2R	DoS	R2L	Precision	Recall
Probe	1279	0	1	0	0.992	0.992
U2R	0	15	0	0	1	0.882
DoS	6	0	117,385	0	0.999	0.999
R2L	4	2	0	331	0.982	1

On the lowest level of the target attribute hierarchy, four classifiers were employed, each for selection of particular subsets of the attacks. We used tree-based classifiers, and each model was trained on the dataset consisting of records of particular attack class. Table 5 summarizes the performance of the models.

Table 5. Performance metrics of particular models used to predict the particular kind of the specific attack types.

	Probe	U2R	DoS	R2L
Overall accuracy	0.991	0.937	0.999	0.989
Precision	0.989	0.927	0.999	0.879
Recall	0.989	0.875	0.999	0.833

4.2.2. Overall Approach Performance

When all particular models in the presented hierarchical detection system were trained, the overall approach was evaluated on the testing subset of KDD 99 dataset. The testing set consisted of records, previously not used during the training of the models, sampled from the KDD 99 dataset. Serialized models were stored and instantiated in the ontology, which was used to navigate in the target attribute hierarchy. For each test record, attack detection model was applied to decide, if the record is an attack or not. To retrieve the attack detection model instance from the ontology, SPARQL query was used. If a record was labelled as an attack, ontology was used to retrieve class taxonomy. Target attribute was specified according to the class hierarchy, and the particular model able to predict the attack types on that level was retrieved and used. Examples of used SPARQL queries are summarized in Table 6.

Table 6. Results of the decision tree classifier for attack prediction.

SPARQL	Action
SELECT ?lname WHERE { ?inst a onto:Connections. ?inst onto:hasModel ?lname	Retrieve the classifier able to predict the attack at the Connection level (decide if the connection is an attack or not)
SELECT ?lname WHERE { ?inst a onto:Attacks. ?inst onto:hasModel ?lname	Select the model for prediction of the attack type

The detection system was evaluated using the same metrics as used in the particular model's evaluation. Table 7. summarizes the system performance—we used the system to predict the type of the attack—and compares the results with other similar approaches [16]. Table 8 presents the confusion matrix of the system and summarizes the precision and recall of particular classes.

Table 7. Performance of the detection system to classify the attack classes.

Classifier	Accuracy	Precision	F-measure	FAR
C4.5	0.969	0.947	0.970	0.005
Random Forests	0.964	0.998	0.986	0.025
ForestPA	0.975	0.998	0.998	0.002
Ensemble model	0.976	0.998	0.998	0.001
Our approach	0.998	0.998	0.998	0.001

Table 8. Performance of the detection system to classify the attack classes.

Ensemble Model	Probe	U2R	DoS	R2L	Normal	Precision	Recall
Probe	1176	0	5	0	7	0.998	0.999
U2R	0	15	0	0	5	0.750	0.937
DoS	4	0	117,547	0	1	0.999	0.999
R2L	3	1	0	346	7	0.969	0.997
Normal	1	0	3	1	48,454	0.999	0.999

4.2.3. Attack Severity Prediction

To demonstrate the usefulness of the ontology, supplementary information related to the attack detection can be provided. For example, in case that the prediction of the concrete attacks is not accurate enough, class prediction can be enhanced by providing the additional aspects of the attack, such as attack severity. Ontology can be used to predict the attack class and the possible severity of the attack. Severity information can be instantiated in the ontology in Severity concept and extracted to data records as a newly created feature. Therefore, we can train a classification model to predict the severity value of the predicted attack. Table 9 describes the severity values of the attack types.

Table 9. The severity of the attacks.

Attack Type	Severity Level
ftp_write	low
guess_passwd	low
spy	low
warezclient	low
warezmaster	low
buffer_overflow	medium
loadmodule	medium
perl	medium
rootkit	medium
phf	medium
imap	medium
multihop	medium
ipsweep	medium
portsweep	medium
nmap	medium
satan	high
back	high
land	high
neptune	high
pod	high
smurf	high
teardrop	high

We used the Severity attribute as the target attribute and trained the Random Forest model to predict the severity level of predicted attack type (predicted by the ensemble model). To train the severity predictor, we used the 10% KDD 99 dataset, split in 70/30 train/test ratio. Table 10. summarizes the prediction of severity types for particular attack classes and evaluation of the accuracy of severity prediction. We ran the severity prediction model as a complementary model to ensemble classifier predicting the attack classes. For each attack class, the severity of the attack was predicted. Confusion table summarizes the severity attribute predictions for particular attack types from the testing set and precision and recall metrics of the severity model. Both, precision and recall are in this case very high. This is the result of a newly created target attribute (Severity level), which has three different values, and particular attacks are well separated within the categories specified by those values. Therefore, standard classifiers perform very well in prediction of such attribute, and the number of missed predictions is minimal.

Table 10. Confusion matrix of the ensemble model for attack severity prediction.

	High	Low	Medium	Precision	Recall
DoS	117,695	0	0		
Probe	443	0	779	0.999	0.999
R2L	0	346	6		
U2R	0	0	20		

5. Discussion and Future Work

The presented approach described the IDS based on a combination of hierarchical ensemble model and knowledge model in the form of an ontology. Both hierarchical and ensemble approaches are relatively popular models used to tackle with the network intrusion detection; the IDS presented in this paper combines both approaches. Hierarchical aspect is considered when the system performs the prediction on a different level of the target attribute generalization. An ensemble model is then employed to detect the attack type to maintain the performance of prediction in major and also minor classes. The ontology then serves to automatically navigate through the target attribute and to select the proper model on the specific level of target generalization. The results proved that the achieved results are comparable to the current state-of-the-art approaches when it comes to performance evaluation using standard metrics. Integration with the knowledge model could be beneficial, as it enables to automate the navigation through the target attribute taxonomy and selection of the appropriate predictive model.

On the other hand, the complexity of the hierarchical and ensemble approach is higher when it comes to training. As each of the models are trained separately, the requirements on training resources and training time are significantly higher than in other compared models. This limitation could be significant when such an approach is expected to be deployed in a real-world environment on dynamic, ever-changing data streams. From this perspective, a very interesting aspect in case of network intrusion detection is the ability to detect the new attacks that are not present in the training set. In real-world scenarios, this corresponds to a specific type of the concept drift (when a new class value appears). To remove this limitation of the standard approaches, either concept drift detection methods or periodic re-training of the models should be employed. Especially in case of ensemble models, early detection of such phenomena is crucial to update the predictive models as soon as possible, so as to not miss any newly appearing attacks. In case of complex ensemble models, the re-training of the partial models could be non-trivial. Moreover, in case of domain knowledge, model is involved (as in the presented IDS model), the update of the model structure is necessary (e.g., addition of the new attack type into the taxonomy).

What could be an also interesting area of research in this field is the deeper integration of the domain-specific knowledge with prediction. The main issue of standard predictive models used in intrusion detection is that the predictions depend on the context of particular events and commonly, such context considers only previous events and their properties. In real-world applications (such as IDS), this context should be expanded with the other, domain knowledge. Predictions then should be the result of information describing the particular event, information about the previous events and information obtained from the domain knowledge model. In this case, the expanded context could be represented by the new, derived features or by the specific expert rules. Such domain expert rules could be used to cover the detection of very rare events (e.g., rare attacks), which could be difficult to extract from the historical data due to a significantly low number of examples. To support such context representation, the knowledge model has to be expanded with more domain concepts and properties, expanding the range of knowledge covered by the ontology.

6. Conclusions

In this paper, we proposed an original symmetrical combination of the knowledge-based approach and machine learning techniques to build a hierarchical intrusion detection system able to support detection of existing types and severity of new types of network attacks. Implemented knowledge model is used in the hierarchical classification mostly to navigate through the target class taxonomy and to select the appropriate model to perform a prediction on the selected level. In case of rare attack types, where the number of available training data is too low, severity prediction model is applied, and the network operator is provided, with information available in the knowledge model about the most probable class of attacks.

The performance of the proposed knowledge-based hierarchical IDS is 0,998 in terms of precision as well as recall and 0,001 in terms of FAR, which outperforms other state-of-the-art approaches presented in Section 2. Moreover, the proposed system is also able to partially cover emerging types of attacks in terms of their predicted severity and additional information stored in the knowledge model. As a result, our knowledge-based approach leads to efficient and information-rich decision support for network operators. In the future work, we plan to extend our approach to make it more dynamic in terms of learning new types of attacks and deciding on the right moments when the available prediction models need to be retrained.

Author Contributions: Conceptualization, J.P.; methodology, M.S. and J.P.; resources, M.S.; software, M.S.; supervision, J.P.; validation, M.S.; writing—original draft, M.S. and J.P. All authors have read and agreed to the published version of the manuscript.

Funding: This work was supported by Slovak Research and Development Agency under the contract No. APVV-16-0213 and by the VEGA project under grant No. 1/0493/16.

Conflicts of Interest: The authors declare no conflict of interest.

References

1. Park, J. Advances in Future Internet and the Industrial Internet of Things. *Symmetry* **2019**, *11*, 244. [CrossRef]
2. Javaid, A.; Niyaz, Q.; Sun, W.; Alam, M. A Deep Learning Approach for Network Intrusion Detection System. In Proceedings of the 9th EAI International Conference on Bio-inspired Information and Communications Technologies (formerly BIONETICS), New York, NY, USA, 3–5 December 2016.
3. Khan, M.A.; Karim, M.d.R.; Kim, Y. A Scalable and Hybrid Intrusion Detection System Based on the Convolutional-LSTM Network. *Symmetry* **2019**, *11*, 583. [CrossRef]
4. Ahmim, A.; Ghoualmi Zine, N. A new hierarchical intrusion detection system based on a binary tree of classifiers. *Inf. Comput. Secur.* **2015**, *23*, 31–57. [CrossRef]
5. Ahmim, A.; Ghoualmi-Zine, N. A New Fast and High Performance Intrusion Detection System. *Int. J. Secur. Appl.* **2013**, *7*, 67–80. [CrossRef]
6. Kevric, J.; Jukic, S.; Subasi, A. An effective combining classifier approach using tree algorithms for network intrusion detection. *Neural Comput. Appl.* **2017**, *28*, 1051–1058. [CrossRef]
7. Srivastav, N.; Challa, R.K. Novel intrusion detection system integrating layered framework with neural network. In Proceedings of the 2013 3rd IEEE International Advance Computing Conference (IACC), Ghaziabad, India, 22–23 February 2013; pp. 682–689.
8. Aljawarneh, S.; Aldwairi, M.; Yassein, M.B. Anomaly-based intrusion detection system through feature selection analysis and building hybrid efficient model. *J. Comput. Sci.* **2018**, *25*, 152–160. [CrossRef]
9. Samrin, R.; Vasumathi, D. Review on anomaly based network intrusion detection system. In Proceedings of the 2017 International Conference on Electrical, Electronics, Communication, Computer, and Optimization Techniques (ICEECCOT), Mysuru, India, 15–16 December 2017; pp. 141–147.
10. Arunadevi, M.; Perumal, S.K. Ontology based approach for network security. In Proceedings of the 2016 International Conference on Advanced Communication Control and Computing Technologies (ICACCCT), Ramanathapuram, India, 25–27 May 2016; pp. 573–578.
11. Salahi, A.; Ansarinia, M. Predicting network attacks using ontology-driven inference. *arXiv* **2013**, arXiv:13040913.

12. Ahmim, A.; Maglaras, L.; Ferrag, M.A.; Derdour, M.; Janicke, H. A novel hierarchical intrusion detection system based on decision tree and rules-based models. *arXiv* **2018**, arXiv:181209059.
13. Sharma, N.; Mukherjee, S. A Novel Multi-Classifier Layered Approach to Improve Minority Attack Detection in IDS. *Procedia Technol.* **2012**, *6*, 913–921. [CrossRef]
14. Ibrahim, H.E.; Badr, S.M.; Shaheen, M.A. Adaptive layered approach using machine learning techniques with gain ratio for intrusion detection systems. *arXiv* **2012**, arXiv:12107650.
15. Gupta, K.K.; Nath, B.; Kotagiri, R. Layered Approach Using Conditional Random Fields for Intrusion Detection. *IEEE Trans. Dependable Secur. Comput.* **2010**, *7*, 35–49. [CrossRef]
16. Zhou, Y.; Cheng, G.; Jiang, S.; Dai, M. An efficient intrusion detection system based on feature selection and ensemble classifier. *arXiv* **2019**, arXiv:190401352.
17. Abdoli, F.; Meibody, N.; Bazoubandi, R. An Attacks Ontology for computer and networks attack. In *Innovations and Advances in Computer Sciences and Engineering*; Sobh, T., Ed.; Springer: Dordrecht, The Netherlands, 2010; pp. 473–476. ISBN 978-90-481-3657-5.
18. Razzaq, A.; Anwar, Z.; Ahmad, H.F.; Latif, K.; Munir, F. Ontology for attack detection: An intelligent approach to web application security. *Comput. Secur.* **2014**, *45*, 124–146. [CrossRef]
19. Zhu, L.; Zhang, Z.; Xia, G.; Jiang, C. Research on Vulnerability Ontology Model. In Proceedings of the 2019 IEEE 8th Joint International Information Technology and Artificial Intelligence Conference (ITAIC), Chongqing, China, 24–26 May 2019; pp. 657–661.
20. Syed, Z.; Padia, A.; Finin, T.; Matthews, L.; Anupam, J. UCO: Unified Cybersecurity Ontology. In Proceedings of the AAAI Workshop on Artificial Intelligence for Cyber Security, Phoenix, Arizona, 12–13 February 2016.
21. Hung, S.-S.; Liu, D.S.-M. A User-centric Intrusion Detection System by Using Ontology Approach. In Proceedings of the 9th Joint Conference on Information Sciences (JCIS), Kaohsiung, Taiwan, 8–9 October 2006; Atlantis Press: Kaohsiung, Taiwan.
22. Abdoli, F.; Kahani, M. Ontology-based distributed intrusion detection system. In Proceedings of the 2009 14th International CSI Computer Conference, Tehran, Iran, 20–21 October 2009; pp. 65–70.
23. Abdoli, F.; Kahani, M. Using Attacks Ontology in Distributed Intrusion Detection System. In *Advances in Computer and Information Sciences and Engineering*; Sobh, T., Ed.; Springer: Dordrecht, The Netherlands, 2008; pp. 153–158. ISBN 978-1-4020-8740-0.
24. More, S.; Matthews, M.; Joshi, A.; Finin, T. A Knowledge-Based Approach to Intrusion Detection Modeling. In Proceedings of the 2012 IEEE Symposium on Security and Privacy Workshops, San Francisco, CA, USA, 24–25 May 2012; pp. 75–81.
25. Karande, H.A.; Gupta, S.S. Ontology based intrusion detection system for web application security. In Proceedings of the 2015 International Conference on Communication Networks (ICCN), Gwalior, India, 19–21 November 2015; pp. 228–232.
26. Can, Ö.; Ünallır, M.O.; Sezer, E.; Bursa, O.; Erdoğdu, B. A semantic web enabled host intrusion detection system. *Int. J. Metadata Semant. Ontol.* **2018**, *13*, 68. [CrossRef]
27. Divekar, A.; Parekh, M.; Savla, V.; Mishra, R.; Shirole, M. Benchmarking datasets for Anomaly-based Network Intrusion Detection: KDD CUP 99 alternatives. In Proceedings of the 2018 IEEE 3rd International Conference on Computing, Communication and Security (ICCCS), Kathmandu, Nepal, 25–27 October 2018; pp. 1–8.
28. Özgür, A.; Erdem, H. A review of KDD99 dataset usage in intrusion detection and machine learning between 2010 and 2015. *PeerJ Preprints* **2016**, *4*, e1954v1.
29. Mavroeidis, V.; Bromander, S. Cyber Threat Intelligence Model: An Evaluation of Taxonomies, Sharing Standards, and Ontologies within Cyber Threat Intelligence. In Proceedings of the 2017 European Intelligence and Security Informatics Conference (EISIC), Athens, Greece, 11–13 September 2017; pp. 91–98.
30. Tavallaee, M.; Bagheri, E.; Lu, W.; Ghorbani, A.A. A detailed analysis of the KDD CUP 99 data set. In Proceedings of the 2009 IEEE Symposium on Computational Intelligence for Security and Defense Applications, Ottawa, ON, Canada, 8–10 July 2009; pp. 1–6.

Article

Malware Classification Using Simhash Encoding and PCA (MCSP)

Young-Man Kwon [1], Jae-Ju An [1], Myung-Jae Lim [1], Seongsoo Cho [2,*] and Won-Mo Gal [1,*]

1 Department of Medical IT, Eulji University, Seongnam 13135, Korea; ymkwon@eulji.ac.kr (Y.-M.K.); 2014162033@stuoffice.eulji.ac.kr (J.-J.A.); lk04@eulji.ac.kr (M.-J.L.)

2 School of Software, Soongsil University, Seoul 06978, Korea

* Correspondence: css3617@gmail.com (S.C.); wongal@eulji.ac.kr (W.-M.G.); Tel.: +82-10-4763-3352 (S.C.)

Received: 25 March 2020; Accepted: 1 May 2020; Published: 19 May 2020

Abstract: Malware is any malicious program that can attack the security of other computer systems for various purposes. The threat of malware has significantly increased in recent years. To protect our computer systems, we need to analyze an executable file to decide whether it is malicious or not. In this paper, we propose two malware classification methods: malware classification using Simhash and PCA (MCSP), and malware classification using Simhash and linear transform (MCSLT). PCA uses the symmetrical covariance matrix. The former method combines Simhash encoding and PCA, and the latter combines Simhash encoding and linear transform layer. To verify the performance of our methods, we compared them with basic malware classification using Simhash and CNN (MCSC) using tanh and relu activation. We used a highly imbalanced dataset with 10,736 samples. As a result, our MCSP method showed the best performance with a maximum accuracy of 98.74% and an average accuracy of 98.59%. It showed an average F1 score of 99.2%. In addition, the MCSLT method showed better performance than MCSC in accuracy and F1 score.

Keywords: malware detection; deep learning; CNN; PCA; dimension reduction; linear transform; Simhash encoding; LSH; symmetrical covariance matrix

---

1. Introduction

Malware is any malicious program that threatens other computer security systems by hacking or stealing personal information for various purposes. There are several methods to check whether a file is malicious or not: static analysis and dynamic analysis [1]. Static analysis is a method that analyzes signature-based information to find a possible defect in an executable file. Such methods primarily rely on pre-processing for pattern recognition because they do not need to run an executable file. On the other hand, dynamic analysis works with behavioral-based information about an executable file to find flaws by executing it in a virtual environment.

Nowadays, both analyses have been combined with deep learning techniques, which has yielded many novel approaches for detecting malware. In particular, static analysis and deep learning techniques do well together [2–4]. In this case, there are three main approaches. The first one uses an executable file as input data for the RNN network [2]. In that paper, the author proposed the RNN network method for classification. The opcode patterns were used as features and a two-stage RNN network using LSTM (long short-term memory) cell was used as a classifier. This resulted in an area under the curve (AUC) of over 0.99 and average AUC of 0.987, respectively. N network. Since the length of every executable file varies, the image file varies. However, because the CNN network takes data of a fixed length, we need to make them the same size. To solve this problem, cropping and hashing methods are used [3,4]. For cropping, Nataraj et al. used the binary code of the executable files [3]. The binary code is mapped to the 8-bit vector, which is the greyscale. As the size of the binary code was different for each file, the size of the image is also different. Thus it needs to be cropped to a fixed size so the image can be used for the CNN classifier. For hashing, Ni et al. used opcode sequences extracted from the executable files [4]. To make the different lengths of opcode sequences a fixed length of vector, they applied a hashing method to the opcode sequence. Because all opcodes of the executable file are mapped to the fixed size of the hash table, this results in the fixed length although the number of opcodes within file is different. We can use them as a (1-dimensional) vector for the PCA or an image (2-dimensional) for the CNN classifier. In the case of latter, we have to reshape them. The third approach is a synthetic way of simultaneously combining the RNN and CNN networks [5]. In that paper, the author used the RNN network to extract features from malicious code and converted the features into an image for training the CNN network.

The main contributions of this paper are: (1) we propose a malware classification method that uses Simhash and PCA (MCSP), which combines Simhash encoding and PCA; (2) we propose a malware classification method that uses Simhash and linear transform (MCSLT), which combines Simhash encoding and linear transform layer. The role of the linear transform layer in MCSLT is to mimic the effect of PCA; (3) we analyzed the cumulative variance according to the N-gram opcode sequence and found the number of principal components; and (4) the performance of the proposed methods were measured and obtained the best results among the compared methods.

The organization of this paper is as follows. In Section 2, we discuss the related work: malware classification using Simhash and CNN (MCSC), n-gram MCSC and PCA. In Section 3, we present and discuss the results of the Simhash encoding and PCA analysis. We present our experimental data, environmental setting and results on the performance of the used methods in Section 4. Finally, we summarize and conclude our overall research in Section 5.

2. Background

2.1. Static, Dynamic and Hybrid Approaches

Static analysis is an approach that analyzes executable files without running them. This method uses the characteristics of the files, such as the string signature, byte sequence n-grams and opcode or opcode distribution, etc. Its advantage is that it is simple to construct and implement because we only need to extract features. However, this method is vulnerable to obfuscation (hiding the original algorithm and data structures) because it does not run the program.

Dynamic analysis methods analyze the behavioral characteristics of executable files while running them. Since a file must be executed to detect malware, it needs a virtual environment such as a simulator or sandbox, etc. It monitors behaviors such as the API call or sign for tainting other files. The advantage of this approach is that it is good for obfuscation. As many malicious programmers now use tricks against detection, this method is of interest because it detects malware based on what execution files actually do. However, it is time-consuming and needs virtual environments.

Nowadays, hybrid approaches, where both static and dynamic methods are combined, have become a subject of intensive research. That is, static and dynamic features are used for detecting malware. In [6], the author used the hybrid method to train their classifier with static and dynamic features. For the static feature, they extracted the opcode sequences using IDA Pro and trained their model with them. For the dynamic feature, they traced each file with IDA Pro using the "tracing" feature. After getting actual mnemonic opcodes, they extracted API calls in them. During training, they trained the model with the opcode sequence and scored it with API calls.

2.2. Simhash, N-Gram, MCSC Image and N-Gram MCSC

Simhash encoding is a local sensitive hash algorithm proposed by Charinikar (2002) and is mainly used for sequence similarity [7]. That is, similar words will have similar hash values.

The N-gram is a method that groups a given sentence or data as sequences of continuous combinations [8]. The size of the combination is defined according to N. When grouping data has N = 1, it is called a "unigram", N = 2 is a "bigram" and N = 3 is a "trigram". N-gram is not only used in natural language processing but also used effectively in malware detection. In [9], the author proposed multi-level big data mining using natural language processing (NLP) and machine learning (ML) techniques for detecting Ransom-ware attacks. N-gram and TF-IDF methods were used for making features and the result is dependent on different sizes of N with an accuracy of 98.59% obtained when N = 3.

The MCSC [4] is an algorithm for malware classification, which converts the executable file into an image and uses it as input data for the CNN network. We refer to this image an MCSC image. The overall procedure is shown in Figure 1.

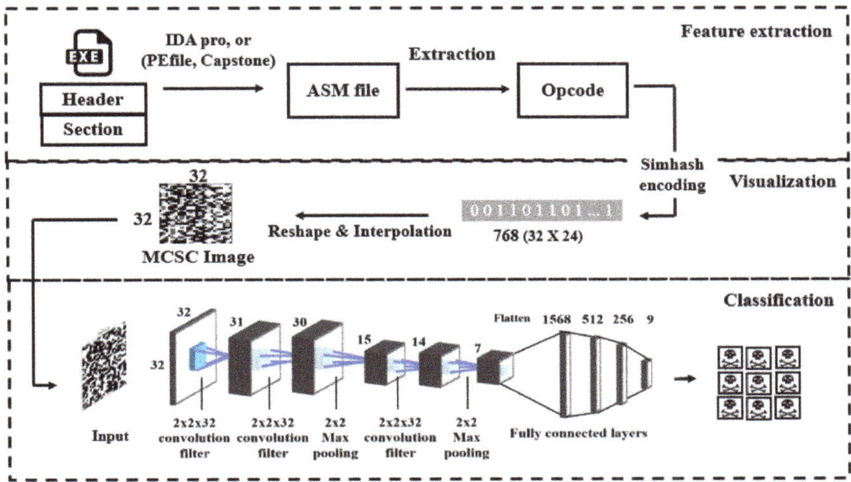

Figure 1. The overall procedure of malware classification using Simhash and CNN (MCSC) algorithm.

The MCSC method has three phases: feature extraction, visualization, and classification. In feature extraction, all opcodes are extracted from the code section in the ASM file, which is made by a disassembler. In the visualization phase, Simhash encoding is undertaken using the opcode sequence. After encoding, the executable file is represented as a fixed length of binary code. Then, the binary code is reshaped into the pre-defined size and interpolated with bilinear interpolation, and the CNN classifier uses it for training.

The original MCSC used several techniques in the step of feature extraction and used SHA-768 simhash encoding, three convolution layers, 2 × 2 filters and 2 max-pooling layers in the classification step. We only used the same classifier of the MCSC method and interpolation techniques except major-block selection. The detailed parameters of the classifier of MCSC are shown in Table 1 because we are concerned with the performance of the classifier. There are three reasons why we used the SHA-768 in our model. First, it showed the best performance in previous MCSC research. Second, we want to compare its performance with our methods using a simple technique (PCA). Lastly, the SHA-768 is composed of SHA-512 and SHA-256, so we can anticipate the effect for the different scales of encoding.

The N-gram MCSC is a method that combines the N-gram and MCSC images [10]. It applies the N-gram method to the original opcode sequence (1-gram) to make an N-gram opcode sequence. We call this the N-gram MCSC image. The N-gram for the opcode sequence reinforces the semantic meaning of the chunked opcode sequence.

Table 1. Detailed parameters for the classification in Figure 1.

Layer (Type)	Output Shape	Param #	Remarks
Conv2d	[−1, 32, 31, 31]	160	
Tanh	[−1, 32, 31, 31]	0	
Conv2d	[−1, 32, 30, 30]	4128	
Tanh	[−1, 32, 30, 30]	0	
MaxPool2d	[−1, 32, 15, 15]	0	
Dropout2d	[−1, 32, 15, 15]	0	
Conv2d	[−1, 32, 14, 14]	4128	
Tanh	[−1, 32, 14, 14]	0	
MaxPool2d	[−1, 32, 7, 7]	0	
Dropout2d	[−1, 32, 7, 7]	0	
Linear	[−1, 512]	803,328	|
Tanh	[−1, 512]	0	|
Linear	[−1, 256]	131,328	| FC classifier
Tanh	[−1, 256]	0	|
Linear	[−1, 9]	2313	|

2.3. Principal Component Analysis (PCA)

PCA is a method for analyzing given data to find principal components through the distribution of the data, where the principal components are the direction vectors with sequentially high variances for the symmetrical covariance matrix of the data [11]. For example, suppose we have two or three-dimensional data. We can find a line that minimizes the distance from each point of the data to one straight line. This line is called as the first principle component of the data. Repeating the same system, we can find the second principle components which are the orthogonal vectors to each other, and each dimension is not correlated. We can obtain two principle components for two-dimensional data and three components for three-dimensional data. PCA can be used for many applications as it applies to many fields. The most popular applications of PCA are dimensionality reduction and exploratory data analysis [12]. Dimension reduction is used to reduce the dimensions of a given dataset to a certain size. For dimensionality reduction, we obtain all the principal components of the given data through PCA, and then the data is projected with the number of principal components that we desire. The number of components is the reduced dimension for the data. Avoiding the curse of dimensionality is one of the main advantages of dimension reduction. Exploratory data analysis is used to understand high-dimensional data using PCA. After finding the principle components, we can see the degree to which the first principal component can explain the data or how many components are needed to represent the dataset in the 90% variances. Also, we are able to visualize high-dimensional data in two or three dimensions.

3. Proposed Methods

3.1. Experimental Dataset

For the classification experiment, we used the Microsoft Challenge dataset [13], which has 9 classes and a total of 10,868 ASM files. With the exception of the files with no opcode after feature extraction, we used 10,736 of these files. Table 2 shows the distribution of malware files. It is highly imbalanced, with the main class being the Kelihos_ver3 virus with almost 3000 files, and a minor class being the Simda virus with 39 files [14].

Table 2. The distribution of malware files.

Virus Name	Number of Files	TYPE
Ramnit	1532	Worm
Lollipop	2470	Adware
Kelihos_ver3	2937	Backdoor
Vundo	447	Trojan
Simda	39	Backdoor
Tracur	732	TrojanDownloader
Kelihos_ver1	387	Backdoor
Obfuscator.ACY	1179	Any kind of obfuscated malware
Gatak	1013	Backdoor

In our dataset, there is Obfuscator.ACY category, which consists of any kind of obfuscated malware such as polymorphism or metamorphism to avoid detection systems. We cannot know what exact techniques are used. If the model can detect properly, we can say that the model is not vulnerable to countermeasures.

In our research, we experimented with Simhash (SHA-768) encoding according to the N-gram opcode sequence. The number of different opcode combinations is used to make N-gram Simhash encoding. This means that the size of the vocabulary dictionary is different. The vocabulary size of the N-gram is shown in Table 3. In the case of the 1-gram, the size of the vocabulary dictionary is 231 words (it is the number of different opcodes).

Table 3. The size of the vocabulary dictionary of N-gram.

N-Gram	The Size of the Vocabulary Dictionary of N-Gram
1-gram	231
2-gram	24,920
3-gram	24,916
4-gram	24,910

3.2. Method 1: Malware Classification Using Simhah Encoding and PCA (MCSP)

Before proposing our methods, we did a principal component analysis (PCA) analysis for Simhash encoding according to N-gram. We analyzed the variance explained by PCA of each Simhash encoding because different vocabulary size is mapped to the constant size (768). The results of the PCA analysis are shown in Table 4 and Figure 2.

In Figure 2, we visualized the cumulative variance according to the number of axes and marked the point for 95% explained variance of the N-gram. As shown in Table 4, the number of PCA axes for 95% explained variance of the 1-gram, 2-gram 3-gram and the 4-gram were 141, 316, 441 and 510, respectively. This indicated that we can reduce each of them by maintaining the 95 % variance of the 768-dimensional Simhash encoding. The number of PCA axes for 90% explained variance for the 1-gram, 2-gram 3-gram and the 4-gram were 53, 162, 278 and 358, respectively.

Table 4. The number of principal component analysis (PCA) axis for explained variance against the total variance.

The Number of PCA Axis\N-Gram	1-Gram	2-Gram	3-Gram	4-Gram
The number of PCA axis for 95% explained variance	141	316	441	510
The number of PCA axis for 90% explained variance	53	162	278	358

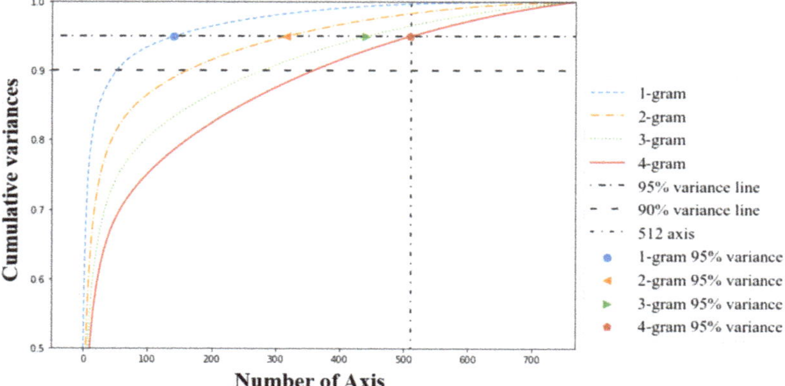

Figure 2. The graph of cumulative variances according to the number of axes.

Based on the above results, we propose a malware classification method using Simhash encoding and PCA (MCSP). The overall procedure of the MCSP method is shown in Figure 3.

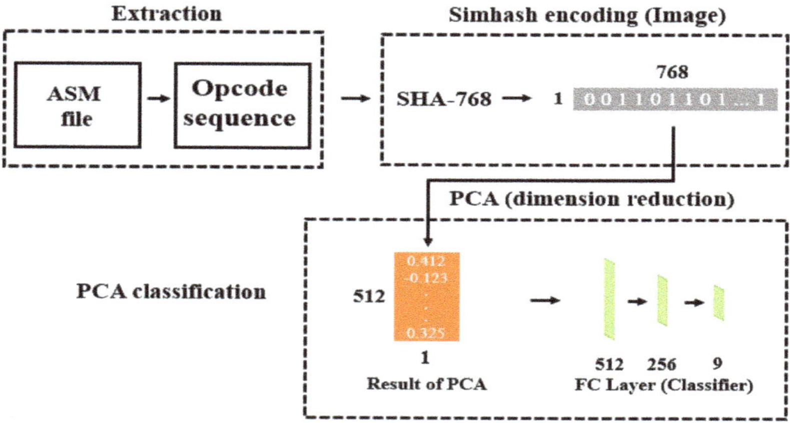

Figure 3. The overall procedure of the malware classification using Simhash and PCA (MCSP) method.

As shown in Figure 3, the MCSP method includes three steps: extraction, Simhash encoding, and PCA classification. In the extraction step, the opcodes sequence is extracted from an asm file. In the Simhash encoding step, the extracted opcode is encoded through Simhash (SHA-768). It is a multi-hash with SHA-512 and SHA-256 [15]. In the PCA classification step, the 768-dimensional Simhash encoding is reduced to 512-dimensional sequences by using the PCA algorithm. The reason for reducing Simhash encoding to 512 dimensions is to ensure that the total variance in the reduced data exceeds at least 95% variance. After that, the transformed sequence is used as an input feature for a fully connected (FC) classifier.

3.3. Method 2: Malware Classification Using Simhash Encoding and Linear Transformation Layer (MCSLT)

We propose another method for malware classification by using Simhash encoding and linear transform layer (MCSLT). It mimics the role of PCA by using a neural network with two layers, which are layers without an activation function. This transforming layer is known as the linear transform (LT) layer. The overall procedure of the MCSLT method is shown in Figure 4.

As can be seen in Figure 4, MCSLT works in three steps: extraction, Simhash encoding and LT classification. The first two steps are the same as the MCSP method. The Simhash encoding is used as the input for the LT layer. We think the first layer of LT reveals another coordinate system and the second layer of LT reduces the dimension. The intention is to let the model learn another coordinate system and reduce the dimension automatically. That is, LT layers are used to find a linear transformation similar to the PCA. The output of the LT layer is used as input features for the same FC layer as the MCSP method.

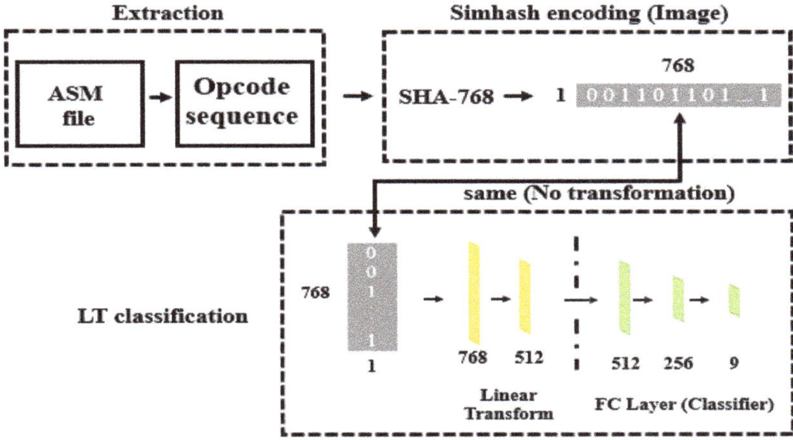

Figure 4. The overall procedure of the malware classification using Simhash and linear transform (MCSLT) method.

4. Experimental Results

We split the dataset into train and test datasets at a proportion of 8:2. Our hardware setting was Intel i7-7600 k, 64 GB memory and 2 Nvidia GeForce GTX 1080 Ti. We implemented the proposed methods with Python 3.6 and Pytorch 1.1.0 in the Jupyter Notebook environment, but we only used 1 gpu for these experiments.

The used Simhash encoding was a multi-hashing method: SHA-768. The SHA-768 consists of SHA-512 and SHA-256. After encoding the N-gram opcode sequence with the SHA-512 and the SHA-256, we combined the outputs. It is the binary sequence of 768 sizes.

To compare the performance of the proposed methods, we used the fully connected (FC) classifier. The overall procedure is shown in Figure 5.

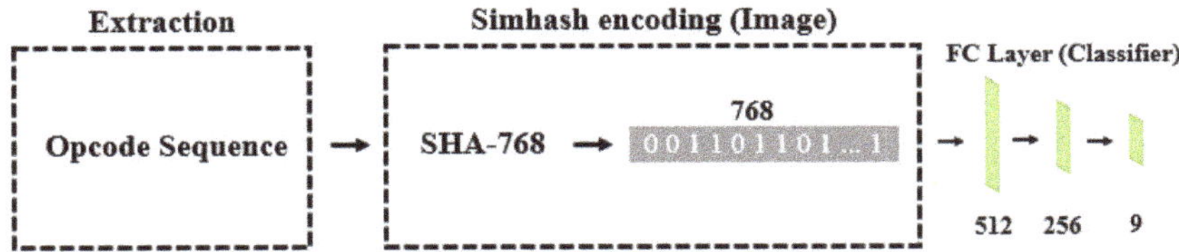

Figure 5. The overall procedure of the fully connected (FC) classifier.

The FC classifier is a simple classifier that uses Simhash encoding as an input feature for the FC layer. The FC layer consists of three dense layers that are 512, 256 and 9 sequentially. Each layer used the relu activation function. This is different to the original MCSC, which used the tanh function for activation, which is symmetric with respect to the original and its derivation is also symmetric. It is only used to measure the basic performance of Simhash encoding.

In the case of MCSC, it reshaped the Simhash encoding into a 32 × 24 image. Then, bi-linear interpolation was applied to make 32 × 32 square images. The interpolated image is used as input data for the CNN structure as shown in Figure 1. The neural network of MCSC consists of 3 convolutional layers using a 2 × 2 filter, tangent hyperbolic activation function and 2 max-pooling layers. This flattens the output of convolution layers and uses the fully connected layers.

In our experiment, we used the basic MCSC method. That is, we did not use any optimization technique in the original study. We also used the two versions of the MCSC method. The first is tanhMCSC, which uses the tanh activation function in the FC classifier and the second is the reluMCSC, which uses the relu activation function in the FC classifier.

The number of parameters for all of the compared methods, is shown in Table 5. The number of parameters for the classifier layers was calculated like this: 133,641 = (512 + 1) × 256 + (256 + 1) × 9. In the case of the FC classifier, we need additional parameters: that is 393,728 = (768 + 1) × 512. Therefore, the total number of parameters is 527,369 as shown in Table 5. The total number of MCSC, MCSP and MCSLT′ parameters are 945,385, 396,297 and 1,477,641, respectively. As a result, MCSP has the fewest parameters with a total of 396,297. However, MCSLT has 1.5 and 2.8 times more parameters than MCSC and FC classifier.

Table 5. The total number of parameters according to the method.

Methods	FC Classifier	MCSC (tanh, relu)	MCSP	MCSLT
Model + Classifier layers	393,728 + 133,641	811,744 + 133,641	262,656 + 133,641	1246976 + 133,641
The total number of parameters	527,369	945,385	396,297	1,477,641

For hyper-parameters of the convolutional neural networks, we assumed the batch size was 64, the epoch was 500, the learning rate was 0.005 and the loss function was cross-entropy. For other models, we used the same batch size, epoch, loss function but the learning rate was 0.0001.

In order to compare the accuracy of the compared methods, we conducted the experiments 30 times for each method. Therefore, we can use a parametric test that is based on the normality of data. The accuracy boxplot for all of the compared methods is shown in Figure 6. In addition, the average and maximum accuracy for them is shown in Table 6.

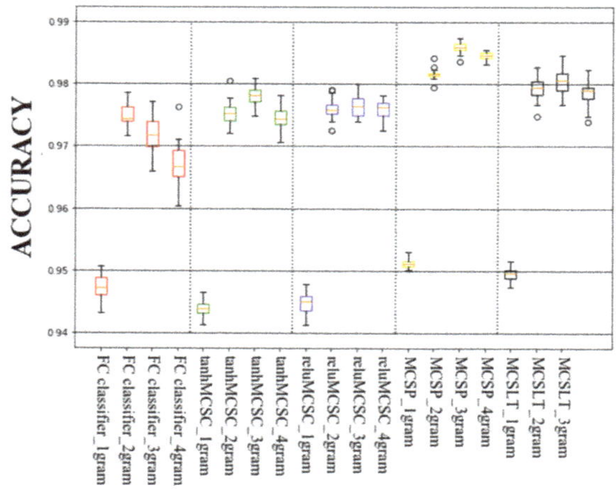

Figure 6. The accuracy boxplot of all compared methods.

Table 6. The accuracy for all compared methods.

N-Gram\Methods		FC Classifier	tanhMCSC	reluMCSC	MCSP	MCSLT
1-gram	Mean	94.73	94.40	94.48	95.13	94.95
	Max	95.07	94.65	94.79	95.30	95.16
	Std	0.0020	0.0012	0.0017	0.0007	0.0011
2-gram	Mean	97.47	97.53	97.60	98.16	97.95
	Max	97.86	98.04	97.91	98.42	98.28
	Std	0.0018	0.0018	0.0014	0.0007	0.0017
3-gram	Mean	97.21	97.80	97.80	98.60	98.04
	Max	97.22	98.09	98.00	98.74	98.46
	Std	0.0031	0.0014	0.0016	0.0008	0.0019
4-gram	Mean	96.71	97.46	97.60	98.47	97.85
	Max	97.63	97.81	97.81	98.56	98.23
	Std	0.0033	0.0018	0.0014	0.0005	0.0020

According to Table 6, the FC classifier showed an average accuracy of 94.73%, 97.47%, 97.21% and 96.71% for 1-gram, 2-gram, 3-gram and 4-gram Simhash encoding, respectively. The tanhMCSC showed 94.40%, 97.53%, 97.80% and 97.46%, respectively. The reluMCSC showed 94.48%, 97.60%, 97.80% and 97.60%, respectively. The MCSP showed 95.13%, 98.16%, 98.60% and 98.47%, and he MCSLT showed 94.95%, 97.95%, 98.04% and 97.85%, respectively.

To examine whether the compared methods were statistically different, we calculated the analysis of variance (ANOVA), except for the FC classifier [16]. There was a significant difference (p value < 0.000) in the 1-gram and 2-gram Simhash encoding. In the post-hoc analysis, we used the Scheffe method with a 95% confidence level. As a result, we found three groups: tanhMCSC and reluMCSC, MCSLT, MCSP (in the order of their performance). There was a significant difference (p value < 0.000) for the 3-gram, which resulted in four groups being identified: reluMCSC, tanhMCSC, MCSLT, MCSP. There was a significant difference (p value < 0.000) for the 4-gram and four groups were identified: tanhMCSC, reluMCSC, MCSLT, MCSP.

With regard to accuracy, the MCSP method showed the best performance in spite using 2.4 times fewer parameters than the MCSC method and 3.7 times fewer parameters than MCSLT. In the case of MCSLT, it performed better than MCSCs for all N-grams. However, it used 1.6 times more parameters than the MCSC method.

We considered the problem of multi-class classification as several binary classification problems for each malware family. So, Precision, Recall and F1 score are common evaluation methods in classification problems. In multi-class classification, there are macro- and micro-average methods [17]. For example, suppose there are k classes, then, micro-averaging for Precision is calculated by the following formula.

$$PRE_{micro} = \frac{TP_1 + \ldots + TP_k}{TP_1 + \ldots + TP_k + FP_1 + \ldots + FP_k} \tag{1}$$

Macro-averaging for Precision is calculated by the following formula.

$$PRE_{macro} = \frac{PRE_1 + \ldots + PRE_k}{k} \tag{2}$$

The F1 score is calculated by the following formula.

$$\text{F1} = \frac{2 * Precision * Reall}{Precision + Reall} \tag{3}$$

In addition to accuracy, we measured the F1 score by using micro-averaging to evaluate the results for 30 run times. These are shown in Figure 7 and Table 7. We found the scores by using the function sklearn.merics.precision_ recall_fscore_support (... , ... , average = 'micro'). In this case, all metrics (Precision, Recall and F1 score) are the same. Therefore, we can also regard the results shown in Figure 7 and Table 7 as Precision or Recall.

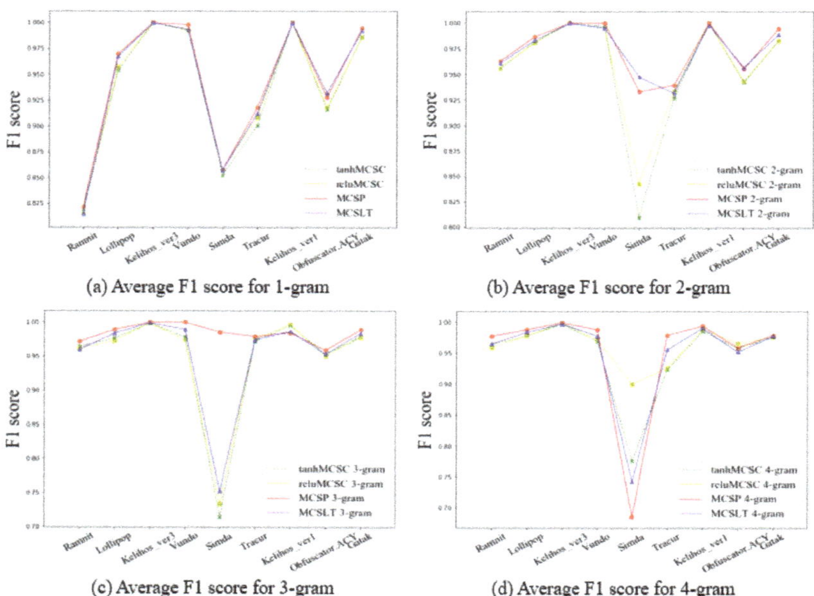

Figure 7. Average F1 score using micro-averaging for all compared methods.

Table 7. The Average F1 score using micro-averaging for all compared methods.

Methods		Ramnit	Lolli pop	Kelihos_ ver3	Vundo	Simda	Tracur	Kelihos_ ver1	Obfuscator. ACY	Gatak	Mean Score	Weighted Mean Score
1-gram	tanhMCSC	0.8193	0.9539	0.9999	0.9924	0.8524	0.9005	1	0.9159	0.9868	0.9357	0.9357
	reluMCSC	0.8417	0.9574	1	0.9924	0.8571	0.9084	1	0.9185	0.9859	0.9372	0.9694
	MCSP	0.8213	0.9699	1	0.9981	0.8571	0.918	1	0.9281	0.9948	0.9431	0.9696
	MCSLT	0.8146	0.9676	1	0.9936	0.8571	0.9126	1	0.9328	0.9924	0.9412	0.9685
2-gram	tanhMCSC	0.9556	0.9814	1	0.9962	0.8095	0.9267	1	0.9425	0.9826	0.9549	0.9549
	reluMCSC	0.9565	0.981	1	0.9951	0.8429	0.9336	1	0.9432	0.9832	0.9595	0.9779
	MCSP	0.963	0.9868	1	1	0.9333	0.9397	1	0.9555	0.9948	0.9748	0.9869
	MCSLT	0.9611	0.9834	1	0.9951	0.9476	0.932	0.9982	0.957	0.989	0.9737	0.9861
3-gram	tanhMCSC	0.9643	0.9767	0.9984	0.9788	0.7143	0.9724	0.9958	0.953	0.979	0.9481	0.9481
	reluMCSC	0.9608	0.9731	0.9988	0.9742	0.7333	0.9756	0.9970	0.9503	0.9782	0.9490	0.9719
	MCSP	0.9717	0.9897	1	1	0.9857	0.9795	0.9848	0.9592	0.9892	0.9844	0.9920
	MCSLT	0.9603	0.9842	0.9997	0.9898	0.7524	0.9749	0.9873	0.9536	0.9844	0.9541	0.9745
4-gram	tanhMCSC	0.9647	0.9782	0.9971	0.9697	0.7762	0.9231	0.9855	0.9583	0.9763	0.9477	0.9477
	reluMCSC	0.9592	0.9784	0.9988	0.9723	0.9000	0.9260	0.9891	0.9667	0.9763	0.963	0.9802
	MCSP	0.9779	0.9886	1	0.9886	0.6857	0.9797	0.9945	0.9595	0.9796	0.9505	0.9716
	MCSLT	0.9658	0.9844	0.998	0.9784	0.7429	0.9562	0.9909	0.9525	0.978	0.9497	0.9718

For the 1-gram, the average micro F1 score of the tanhMCSC method were 81.93%, 95.39%, 99.99%, 99.24%, 85.24%, 90.05% 100%, 91.59% and 91.85% according to malware classes ranging from Ramnit through to Gatak. The mean F1 score of the tanhMCSC method was 93.57%. For the 1-gram Simhash encoding, the mean F1 score for the four compared methods were 93.57%, 93.72%, 94.31%, and 94.12%, respectively. In Table 7, we added the last column (shaded column), which shows the weighted macro F1 score. We found this score by using the function sklearn.merics.f1_score(... , ... , average = 'weighted'). We plotted the values of the last two columns in Table 7, the mean F1 score and the mean weighted F1 score, according to N-gram for the compared methods in Figure 8.

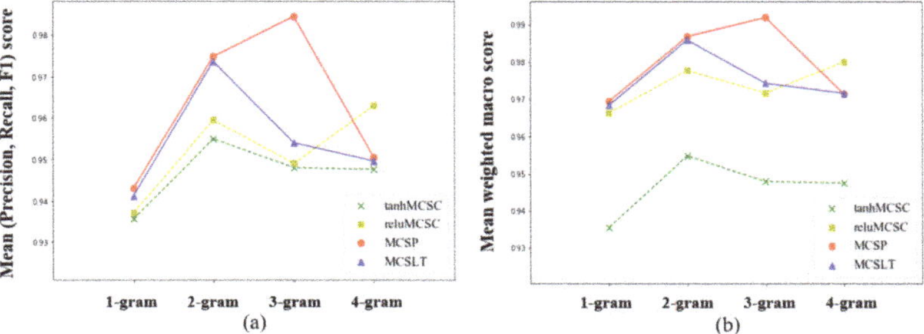

Figure 8. Mean (Precision, Recall, F1) score (**a**) and mean weighted macro score (**b**) for compared methods.

Our data is highly imbalanced, therefore, the weighted macro F1 score seemed to be the right metric to evaluate. Our MCSP and MCSLT showed better performance than the tanhMCSC method for all N-gram Simhash encoding. Our methods performed better than the reluMCSC method for the 1-gram, 2-gram and 3-gram. However, they showed a lesser performance for the 4-gram.

From the viewpoint of the weighted macro F1 score, MCSP showed the best performance with a value of 99.2% for the 3-gram. The MCSLT model showed the best performance with a value of 98.61% for the 2-gram, although, this value is less than the MCSP method. In the case of reluMCSC, the model showed the best performance with a value of 98.02% for the 4-gram.

5. Conclusions

In this paper, we proposed two malware classification methods known as the MCSP and MCSLT methods. The MCSP method classifies malware using Simhash encoding and PCA. It encoded the opcode sequences of ASM files to Simhash encoding and applied the PCA method to them. The MCSLT method applies the LT layer to Simhash encoding. The LT layer consists of two fully connected layers that mimic the linear transformation, which is similar to PCA.

The Microsoft Challenge public dataset was used to evaluate the performance of tanhMCSC, reluMCSC and the two proposed methods. With regard to the number of parameters for each model, our MCSP model is efficient because it has a simple structure compared to MCSCs that use CNN and it has fewer parameters than them. However, our MCSLT model has 1.6 times more parameters than MCSCs.

With regard to performance, we measured accuracy and F1 score using a micro-average and weighted macro-average. We conducted 30 experiments for Simhash encodings that are derived from the 1-gram, 2-gram, 3-gram and 4-gram opcode sequences. As a result of the experiments, the MCSP had the best maximum accuracy of 98.74% and an average accuracy of 98.58% for the 3-gram Simhash encoding. From the viewpoint of the F1 score using the weighted macro-average, the MCSP and MCSLT showed better performance than the tanhMCSC method for all N-gram Simhash encoding. Our proposed methods showed better performance than the reluMCSC method for the 1-gram, 2-gram and 3-gram. However, they showed less performance for the 4-gram.

The strengths of this paper are that we tackled a subject of current intensive research, the algorithms are very simple, and our methods have computational advantages compared to the existing methods. However, the main limitation of the study is that we did not use the label information for data.

Author Contributions: Conceptualization, W.-M.G., S.C., and Y.-M.K.; Methodology, S.C. and Y.-M.K.; Software, J.-J.A. and M.-J.L.; Validation, W.-M.G. and S.C.; Formal Analysis, W.-M.G. and S.C.; Investigation, J.-J.A., M.-J.L. and Y.-M.K.; Resources, J.-J.A. and Y.-M.K.; Data Curation, J.-J.A. and S.C.; Writing—original draft preparation, J.-J.A., Myung-Jae Lim and Y.-M.K.; Writing—review and editing, M.-J.L. and S.C.; Visualization, J.-J.A. and Y.-M.K.; Supervision, Y.-M.K.; Project administration, Y.-M.K.; Funding acquisition, Y.-M.K. All authors have read and agreed to the published version of the manuscript.

Funding: This research was funded by R.O.K National Research Foundation, grant number NRF-2017R1D1A1B03036372 in 2020.

Conflicts of Interest: The authors declare no conflict of interest.

References

1. Gandotra, E.; Bansal, D.; Sofat, S. Malware analysis and classification: A survey. *J. Inf. Secur.* **2014**, *2014*, 44440. [CrossRef]
2. Lu, R. Malware Detection with LSTM using Opcode Language. *arXiv* **2019**, arXiv:1906.04593.
3. Nataraj, L.; Yegneswaran, V.; Porras, P.; Zhang, J. A comparative assessment of malware classification using binary texture analysis and dynamic analysis. Available online: http://www.csl.sri.com/users/vinod/papers/aisec17-nataraj.pdf (accessed on 2 March 2020).
4. Ni, S.; Qian, Q.; Zhang, R. Malware identification using visualization images and deep learning. *Comput. Secur.* **2018**, *77*, 871–885. [CrossRef]
5. Sun, G.; Qian, Q. Deep learning and visualization for identifying malware families. *IEEE Trans. Dependable Secure Comput.* **2018**. [CrossRef]
6. Damodaran, A.; Di Troia, F.; Visaggio, C.A.; Austin, T.H.; Stamp, M. A comparison of static, dynamic, and hybrid analysis for malware detection. *J. Comput. Virol. Hacking Tech.* **2017**, *13*, 1–12. [CrossRef]
7. Charikar, M.S. Similarity estimation techniques from rounding algorithms. In Proceedings of the thiry-fourth annual ACM symposium on Theory of computing, Montreal, QC, Canada, 19–20 May 2020; pp. 380–388.
8. Witten, I.H.; Frank, E. Data mining: Practical machine learning tools and techniques with Java implementations. *Acm Sigmod Rec.* **2002**, *31*, 76–77. [CrossRef]
9. Poudyal, S.; Dasgupta, D.; Akhtar, Z.; Gupta, K. A multi-level ransomware detection framework using natural language processing and machine learning. In Proceedings of the 14th International Conference on Malicious and Unwanted Software" MALCON, Nantucket, MA, USA, 2–4 October 2019.
10. Lim, M.J.; Kwon, Y.M. Efficient algorithm for malware classification: N-gram MCSC. *Int. J. Comput. Digit. Syst.* **2020**, *9*, 179–185.
11. Wold, S.; Esbensen, K.; Geladi, P. Principal component analysis. *Chemom. Intell. Lab. Syst.* **1987**, *2*, 37–52. [CrossRef]
12. Zhang, T.; Yang, B. Big data dimension reduction using PCA. In Proceedings of the IEEE International Conference on Smart Cloud (SmartCloud), New York, NY, USA, 18–20 November 2016; pp. 152–157.
13. Ronen, R.; Radu, M.; Feuerstein, C.; Yom-Tov, E.; Ahmadi, M. Microsoft malware classification challenge. *arXiv* **2018**, arXiv:1802.10135.
14. Krawczyk, B. Learning from imbalanced data: Open challenges and future directions. *Prog. Artif. Intell.* **2016**, *5*, 221–232. [CrossRef]
15. Dang, Q.H. *Secure hash standard*; (No. Federal Inf. Process. Stds.(NIST FIPS)-180-4); National Institute of Standards and Technology: Gaithersburg, MD, USA, 2015.
16. Miller, R.G., Jr. *Beyond ANOVA: Basics of applied statistics*; CRC Press: Boca Raton, FL, USA, 1997.
17. Van Asch, V. Macro-and micro-averaged evaluation measures. Available online: https://pdfs.semanticscholar.org/1d10/6a2730801b6210a67f7622e4d192bb309303.pdf (accessed on 2 February 2020).

MDPI
St. Alban-Anlage 66
4052 Basel
Switzerland
Tel. +41 61 683 77 34
Fax +41 61 302 89 18
www.mdpi.com

Symmetry Editorial Office
E-mail: symmetry@mdpi.com
www.mdpi.com/journal/symmetry

www.ingramcontent.com/pod-product-compliance
Lightning Source LLC
LaVergne TN
LVHW071004170726
843515LV00006B/1066

* 9 7 8 3 0 3 9 3 6 6 4 2 2 *